Charged Particle Optics Theory

OPTICAL SCIENCES AND APPLICATIONS OF LIGHT
Series Editor
James C. Wyant

*Please visit our website **www.crcpress.com** for a full list of titles*

Timothy R. Groves

SUNY Polytechnic Institute, State University of New York, USA

Charged Particle Optics Theory

An Introduction

CRC Press
Taylor & Francis Group
Boca Raton   London   New York

CRC Press is an imprint of the
Taylor & Francis Group, an **informa** business

Cover picture: A monatomically thick graphene layer, viewed in a Nion UltraSTEM200 aberration-corrected scanning transmission electron microscope (STEM) located at the Oak Ridge National Laboratory. The microscope was operated at 60 KV in annular dark field (ADF) mode. The red regions represent individual carbon atoms with nearest-neighbor spacing of 0.14 nm. The red regions represent the highest scattered intensity, and the blue regions represent the lowest scattered intensity. The image is a sum of intensities of 350 separate areas of a single graphene specimen, each having 128×128 pixels. Each individual scan is translated so as to maximize the cross-correlation function between separate scans. This greatly reduces noise in the composite image. The original image and specimen were prepared by Dr. Florence Nelson, and the correlation analysis was performed by Jennifer Passage. Both are affiliated with the College of Nanoscale Science and Engineering, SUNY Polytechnic Institute, Albany, NY.

The electron microscopy was guided by Dr. Juan-Carlos Idrobo at ORNL in the Shared Research Equipment (ShaRE) User facility, sponsored by the Office of Basic Energy Sciences, U.S. Department of Energy.

CRC Press
Taylor & Francis Group
6000 Broken Sound Parkway NW, Suite 300
Boca Raton, FL 33487-2742

© 2015 by Taylor & Francis Group, LLC
CRC Press is an imprint of Taylor & Francis Group, an Informa business

No claim to original U.S. Government works

Printed on acid-free paper
Version Date: 20150323

International Standard Book Number-13: 978-1-4822-2994-3 (Hardback)
International Standard Book Number-13: 978-0-367-37796-0 (Paperback)

Library of Congress Cataloging-in-Publication Data

Groves, Timothy R.
    Charged particle optics theory : an introduction / author, Timothy R. Groves.
        pages cm -- (Optical sciences and applications of light)
    Includes bibliographical references and index.
    ISBN 978-1-4822-2994-3 (hardcover : alk. paper) 1. Particle beams. 2. Electron optics. 3. Beam optics. I. Title.

QC793.3.B4G76 2015
539.7'3--dc23
                                                                        2015001451

Visit the Taylor & Francis Web site at
http://www.taylorandfrancis.com

and the CRC Press Web site at
http://www.crcpress.com

Contents

Preface

This book is a theoretical introduction to the optics of charged particle beams. The purpose is to identify the most important ideas and derive them mathematically from first principles of physics. It is a teaching document, intended for an audience of students in the broad sense. As a science book, it focuses on basic principles in a connected way. It is intended for the intelligent non-expert who is comfortable with calculus at an advanced undergraduate level. Experts, including experimentalists, instrument designers, and instrument users, will also find it to be a convenient reference for understanding the theoretical origins of the subject.

Enormous experimental progress has been made in recent years, culminating in commercial availability of aberration-corrected transmission electron microscopes with resolution below 0.1 nm, energy analyzers with resolution in the meV range, and gas field ion microscopes with resolution below 1.0 nm, to name a few examples. These innovations are built upon the ongoing efforts of pioneers over the past decades. These advances enable an ever-growing array of applications at the atomic scale of dimensions. Unfortunately, the underlying theory can appear arcane and baffling to someone who is new to the field. One cannot possibly understand aberration correction without first having a firm grasp on optics in the paraxial approximation, and the origin of the primary aberrations, for example.

This book is intended to convey an intuitive understanding of the basics, as opposed to presenting a comprehensive compendium of the detailed subject. It is meant to be logical, with each step following directly from the preceding step, insofar as this is possible. For this reason, it is highly recommended that the reader adhere to the logical sequence, and make the effort to follow the mathematical steps along the way. Problems are included to amplify and fill in the theoretical details, and to provide practical examples.

Many excellent books have been written over the years on this general topic. Indeed we have attempted to include these in the references. As the subject has matured, the various topics have been treated in increasing detail and precision in the literature. In order to present an up-to-date review of the subject, it is common practice for authors to present the main results only, referring the reader to a list of earlier references for detailed derivations and justifications. The methodology here is quite different. All of the ideas presented are derived from first principles of physics. In some instances this excludes the most recent detailed and precise results of others. The idea is to convey an intuitive scientific feel for the subject.

It is standard practice in physics research that, if a particular problem cannot be solved, a related problem is identified which *can* be solved. This inevitably involves approximation. This approach is used here in several instances, most notably in the descriptions of particle scattering and electron emission from solids.

We begin with a general introduction in Chapter 1, consisting of a non-mathematical survey of the optical nature of a charged particle beam. A number of practical systems are described that highlight the enormous breadth and depth of present-day applications.

Next, Chapter 2 describes geometrical optics. This begins with a review of relativistic classical mechanics for the motion of a single particle with general charge q and rest mass m. Based on this, the principles underlying geometrical optics are then derived, including a prescription for solving for the *ray* path, which is the physical path taken by a single particle. This chapter is completely accurate with respect to the special theory of relativity. Interestingly, this adds no significant complexity over the historical non-relativistic treatments, but does lead to a more accurate mathematical description. We therefore keep everything relativistically correct to the extent possible.

Chapter 3 describes wave optics. We begin with a review of quantum mechanics, limited to only those ideas that impact the motion of a single charged particle. We begin with the non-relativistic approximation and Schrödinger's equation. Relativity is introduced later in the form of the Klein–Gordon equation. This skirts a rigorous treatment of spin, but keeps things from becoming too abstract, while producing a practical result. The discussion culminates with the quantum mechanical solution for the propagation of the single-particle wave function in a general electromagnetic potential. The correspondence between wave optics and geometrical optics in the classical limit emerges naturally from this discussion.

We then discuss diffraction and interference, starting with Huygens' principle, and proceeding through the scalar Helmholtz equation, the Huygens–Fresnel relation, the Fresnel approximation, and the Fraunhofer approximation. Next we discuss a number of useful examples, including formation of an image and a diffraction pattern, the general optical transformation from object to image, and the fundamental relationship between diffraction and Heisenberg's uncertainty principle.

Chapter 4 describes the two-body scattering problem, which is basic to the interaction of a fast charged particle with matter. Most of the relevant information about the scattering process is contained in the scattering cross section, which is derived first in the classical approximation, and then in the quantum mechanically. Chapter 5 describes electron emission as a practical consequence of quantum mechanics. Finally, the appendices contain two essential mathematical topics, which are repeatedly referred to in the main text.

All useful information about the motion of a single charged particle is contained in the integral of the classical Lagrangian function between two arbitrary points in time. This integral is known historically as *Hamilton's principal function*, and alternatively as the *eikonal* function. The actual path taken by the particle, chosen among a multiplicity of mathematically possible paths, is the path for which this integral has an extremum. In the important special case where the general electromagnetic potential has no explicit time dependence, the action integral reduces to a line integral of the canonical momentum component along the ray path. This is a considerable simplification in problems where one is only interested in the spatial coordinates of a ray, without the need to know the arrival time at any given point. The extremum condition is generally known as the *principle of least action*, which is expressible in concise and precise mathematical terms.

In quantum mechanics all relevant information about the motion of a single particle is contained in the wave function, for which the same action integral in units of Planck's constant $\hbar$ is the phase. It follows that all possible paths in the immediate vicinity of the classical path interfere constructively. The classical path is thus the path that maximizes the probability. This clarifies the particle–wave duality in concise and elegant mathematical terms. A close analogy exists between Fermat's principle of light optics and the principle of least action for a charged particle. The analogy between light optics and charged particle optics is deep, and is manifested in quite practical ways, including diffraction and interference. These ideas are derived mathematically from first principles.

The literature of this mature field is extensive. Several books are of particular interest. The three-volume set by Hawkes and Kasper [43, 44, 45] describes the main principles in precise and comprehensive detail, with reference to the work of many authors over the decades. There is arguably no better review of the enormous body of work that brought the field to its present state. *Geometrical Charged-Particle Optics* by Rose [75] is both general and comprehensive in its mathematical description of systems with general curvilinear axes. This includes the straight optic axis and axial symmetry as special cases, and also includes the theory of correction of geometrical aberrations. Correction of spherical aberration in electron microscopes is described in a detailed, up-to-date way. This book also derives the main ideas from first principles of physics. The book *Handbook of Charged Particle Optics*, edited by Orloff [67], describes a variety of experimental and theoretical topics in a way that is accessible to readers with a range of experience. It also describes correction of spherical aberration.

The present book complements these in several important ways. It is an introductory textbook that prepares the student to tackle the detailed and comprehensive literature. It proceeds from first principles of physics in a structured way, including geometrical optics (classical mechanics), wave optics (quantum mechnics), and the correspondence between them. Finally, it includes several topics not normally included in other books on charged particle optics, but that are essential to practical systems. These include a first-principles theory of Coulomb interaction in charged particle beams, particle scattering by materials, and electron emission from materials.

Chapter 1

Introduction: The optical nature of a charged particle beam

Modern physics teaches that all matter is made of particles which interact with one another. Every particle is characterized by its intrinsic charge, mass, and spin. These quantities govern all interactions which a particle can have. For example, an atom consists of a cloud of negatively charged electrons orbiting a compact, positively charged nucleus. The establishment of this fact in quantitative terms has a fascinating history. It originates with the early hypothesis of Democritas, proceeds through the origins of quantitative chemistry in the seventeenth century, and culminates with the elucidation of quantum mechanics in the twentieth century. Only during the last few decades has it become possible to capture an actual image of a single atom.

Atoms are charge-neutral in their normal state, with the positive charge of the nucleus precisely offset by negative charge of the orbiting electrons. By bombarding an atom with a beam of light or charged particles, it is possible to remove one or more electrons from an atom or molecule. This forms a positively charged ion. Under special circumstances it is also possible to add electrons to form a negatively charged ion. Electric and magnetic fields act on the intrinsic charge of electrons and ions through the force known as the Lorentz force, after the physicist who first identified it in the nineteenth century. By bombarding with a very high energy beam, the atomic nucleus can dissociate into its constituent elementary particles. This is the mechanism by which a high energy particle accelerator is used to probe the fundamental makeup of matter.

Many examples of free charged particles exist in nature. Energetic ions appear as cosmic rays which pervade interstellar space, and bombard the earth's atmosphere in large numbers. A large variety of subnuclear particles are produced in high energy particle accelerators. Many of these also appear as cosmic rays. The beam inside an electron microscope or a cathode ray tube consists of free, energetic electrons in a vacuum. Indeed, it is not difficult to form a beam of charged particles in a vacuum by making use of the intrinsic properties of matter, together with electric and magnetic fields to focus and steer the beam.

According to the laws of classical physics, a single charged particle traces out a path of motion under the influence of electric and magnetic fields. A collection of many particles emitted from a source, each with its own trajectory, form a beam.

Two common sources are shown schematically in Figure 1.1. In (a) a hot tungsten wire at the top of the figure, with a temperature of about 2000 degrees Kelvin is placed opposite a planar electrode called the anode. The anode is typically electrically grounded. Electrons are spontaneously emitted from the hot wire by the process of thermionic emission. By means of an external power supply, the tungsten wire is elevated to a negative voltage which can be anywhere between a few volts to a few millions of volts relative to the anode. This voltage is called the accelerating voltage, because the resulting electric field accelerates the particles. This forms a beam, which is analogous in several fundamental ways to a beam of light. Each trajectory in the figure corresponds to the path of a single charged particle.

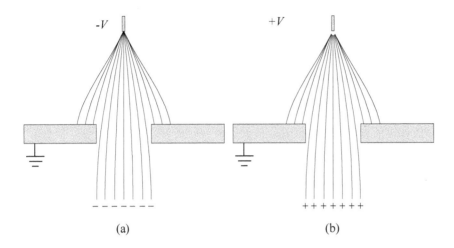

Figure 1.1: (a) electron source, and (b) positive ion source.

In (b) a tungsten wire is formed into a very sharp tip. The tip is elevated to a positive voltage, typically a few thousand to a few tens of thousands of volts, relative to the planar electrode. A small amount of helium gas is admitted into the system. Helium atoms diffuse toward the vicinity of the tip, where they are ionized in the very high electric field. This is known as a gas field ionization source. Ion sources of other chemical species exist as well. Practically any material which can be ionized can be used to form an ion beam. This enables a rich variety of species of ion beams to be formed.

In all cases, an electric field accelerates the charged particles. Each particle acquires an energy equal to its charge times the accelerating voltage. A natural unit of energy is the *electron-Volt*, abbreviated as eV. It is the energy which a particle with one electronic charge acquires when accelerated through one volt. The beam energy is thus easily tuned to almost any desired value by simply controlling the accelerating voltage. This turns out to have considerable practical utility. Practical charged particle beams range in energy from a few eV to about fourteen trillion eV. This is the design energy of the Large Hadron Collider (LHC) at CERN, the world's most energetic particle accelerator, located on the France-Switzerland border. Incidentally, the beam must be in a vacuum chamber in all useful particle beam instruments, since the particles would immediately be absorbed in air at normal atmospheric pressure, regardless of their energy.

A charged particle beam is conceptually similar in many respects to a beam of light. It is therefore interesting to think about charged particle optics in an analogous way to light optics. This forms a central theme in the present study. For example, electric and magnetic fields can be configured to form a lens, which focuses the charged particle beam. An example of a magnetic lens is shown schematically in Figure 1.2. A current-carrying solenoid is depicted in the figure by the two rectangles, which represent the cross sec-

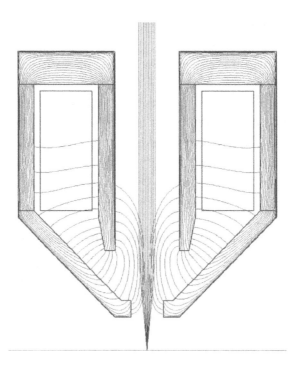

Figure 1.2: Magnetic focusing of a beam of electrons.

tion. The solenoid is surrounded by a shroud of soft iron, which concentrates the magnetic field. The magnetic field lines bulge into the region of the electron beam, which is incident from the top of the figure. The beam is focused to a small probe at the target plane, shown at the bottom of the figure. Such an arrangement is used in a scanning electron microscope. The magnetic field lines and the electron trajectories are generated in a computer simulation by MEBS, Ltd. [63]. The beam path is 100 mm long in the figure, the beam energy is 10 KeV, and the solenoid carries 550 ampere-turns. In reality, the electrons spiral around the central optic axis. The figure is plotted in a coordinate system which rotates about the axis with the beam, so that the trajectories appear not to rotate. This is for clarity.

An example of an electrostatic lens is shown schematically in Figure 1.3. Electrons are emitted from a heated flat surface at zero

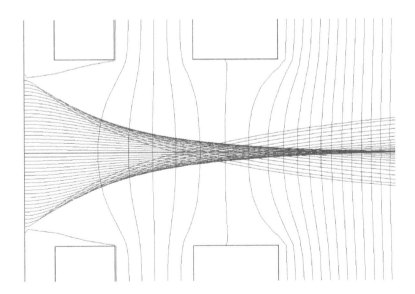

Figure 1.3: Electrostatic acceleration and focusing of a beam of electrons.

volts relative potential on the left of the figure, and are accelerated to the right. An aperture at −10 volts forms a grid to control the total beam current. A second aperture at +600 volts forms an extraction field for emission. Finally, a high voltage electrode at +18,000 volts is located far to the right, out of the figure. The apertures both have diameter 0.6 mm, and the other dimensions in the figure scale proportionally. The curved equipotentials penetrate the space occupied by the beam, and are separated by 100 volts in the figure. These equipotentials can be regarded as forming a lens, which focuses the beam to a crossover at the right of the figure. Such an arrangement is used in a cathode ray tube. The electrostatic equipotential surfaces and the electron trajectories are generated in a computer simulation by MEBS, Ltd. [63].

In addition to focusing a beam to a pointlike spot, a lens can also be used to form a magnified image of an extended object. This is shown schematically in Figure 1.4. Every object point in

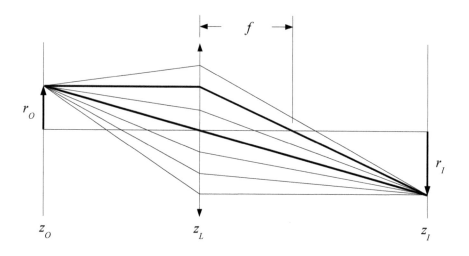

Figure 1.4: Imaging an off-axis point by a lens.

the plane located at z_O emits a cone of rays into the lens at plane z_L. A particular object point is located a vertical distance r_O from the central axis in the figure. A ray which is emitted in a direction parallel with the central axis is deflected by the lens, and intersects the central axis at the *focal point* located an axial distance f from the lens. A second ray passes through the center of the lens, and is undeflected. These two rays are depicted as bold lines in the figure. They intersect in a distant plane located at z_I, at a point which is located at a distance r_I from the central axis. In fact, all rays emitted from this object point, regardless of their angle of emission, are focused by the lens to the same image point. Since all rays converge to a single point, it is apparent that a one-to-one mapping of the object point into an image point exists. In order for this to happen, each ray must experience a change in slope which is proportional to the distance from the central axis. This is the remarkable focusing action of an ideal lens.

Since this works for *any* point in the object plane z_O, we deduce that all object points are imaged simultaneously, each to a unique point in the image plane. This is the mechanism by which a magnified image of an extended object is formed. The negative of the ratio of r_I to r_O is called the *magnification* of the image relative to the object. By convention, the magnification is negative in this case, because the image is inverted relative to the object. By performing the construction in Figure 1.4 for multiple object points r_O, it is easy to convince oneself that this magnification is the same for all object points. The magnification depends only on the relative positions of the object plane z_O and the lens plane z_L, and on the focal length f. The smaller the focal length f, the more the rays are deflected, and the stronger is the lens. The focal length is the same for all object points r_O. For a charged particle beam, the focal length also depends on the particle energy. The higher the particle energy, the longer is the focal length. This is a direct result of the fact that a faster particle spends less time in the lens field, and is therefore deflected less than a slower particle.

The construction in Figure 1.4 works for both charged particles and light. Many striking similarities exist between light optics and charged particle optics. In both cases, no optical system is capable of forming a perfect image. Blur and distortion are always present to some degree. These imperfections are called *aberrations*. An important example is the so-called spherical aberration, in which the outermost rays are focused more strongly than the innermost rays. As a result, the beam is not focused to a point, but rather is blurred. This is readily apparent in Figure 1.3. Spherical aberration occurs in light optical lenses as well as charged particle lenses. In light optics, it arises from the fact that ordinary lenses have spherical surfaces, hence the name spherical aberration. It is substantially corrected in light optics by grinding the lens surfaces to a particular aspherical shape. It is not possible to shape the electric and magnetic fields of a charged particle lens in an analogous way, because the fields always obey Maxwell's equations. Significant progress has been made over the last two decades in correcting the aberrations of charged particle lenses. The details are beyond the scope of this study. The reader is referred to two excellent references by Rose [75] and by Krivanek, et. al. [54] for precise details. Indeed, it is hoped that the present study will provide the background needed to approach this advanced topic expeditiously.

It is apparent from Figure 1.3 that the innermost rays close to the central axis are less aberrated than the outermost rays. Selecting the inner rays and blocking the outer rays would improve the quality of the focusing. This suggests a simple way of mitigating the effect of the spherical aberration for a given optical system, namely, by using an aperture to admit the inner rays, while blocking the outer rays. Conceptually, one could add an aperture in the lens plane of Figure 1.4, thus limiting the cone of rays. A convenient measure of the constriction is given by the index of refraction times the sine of the angle which the extreme ray makes with the central axis at an object point on the axis. This product is known as the *numerical aperture*. The larger the aberration, the smaller the numerical aperture must be to obtain the desired image quality. In fact, the size of the numerical aperture can be used as a useful estimate of the quality of the optical system. In practice, the numerical aperture is typically in the range of 0.3 to 1.3 for light optical lenses, and 0.001 to 0.1 for charged particle lenses, in order to achieve optimal imaging conditions. This expresses the fact that charged particle lenses have significantly worse aberration than light optical lenses.

Classical mechanics regards a single particle as a hypothetical point, with the position and velocity known in principle at any given instant in time. In reality, a single particle also behaves like a wave. The wavelength is is equal to Planck's constant h divided by the particle momentum, where $h = 6.6261 \times 10^{-34} Joule \cdot sec$. A faster particle thus has a shorter wavelength than a slower particle. This so-called wave-particle duality is a hallmark of quantum mechanics, which is a more accurate description of nature than classical mechanics on the atomic and subatomic scale of dimensions. Classical mechanics is sufficiently accurate for many purposes, however, so it is worth retaining. Quantum mechanics has a very specific correspondence with classical mechanics for a charged particle in the limit of high energy. This will prove to be a central theme in the present study.

Quantum mechanics teaches that the absolute square of the wave amplitude is equal to the probability that a single measurement finds the particle at a given position at any given instant in time. Because this probability is described by a propagating wave, it is not possible to know the position and momentum simultaneously with perfect precision. This is known as the Heisenberg uncertainty principle, after the physicist who first elucidated it in the 1920s. A remarkable consequence of quantum mechanics, and one which may appear counterintuitive at first, is that a single particle can be described by two or more waves which interfere constructively or destructively with one another. Each wave corresponds to a particular alternative path of motion of the particle, where the actual path of motion is fundamentally unknowable. For example, it is impossible to know which path in Figure 1.4 is the actual path taken by the charged particle. Each possible path can be described by a separate wave, where all of the waves corresponding to the different paths propagate coherently, with a particular phase relationship to one another. They all interfere at the image plane to cause a blurred spot (not depicted in the figure).

This interference is intimately related to diffraction, which results from the propagation, spreading, and interference of waves.

Diffraction is familiar in light optics. For example, it imposes a fundamental limit on the resolution of a microscope. Because of diffraction, it is not possible in a conventional microscope to resolve any object which is appreciably smaller than the wavelength. This turns out to be true for both a light microscope and an electron microscope. It is another example of the close analogy that exists between charged particle optics and light optics. Since the wavelength of a fast charged particle is much smaller than that of visible light, it is expected that the resolving power of a charged particle microscope should be much better than a light microscope. This is indeed verified in practice. A modern electron microscope can resolve a single atom, a feat which is in no way possible with visible light.

Charged particles interact strongly with matter. This forms the basis of many useful instruments. For example, a fast electron can be scattered by an atomic nucleus of the target material, with little energy transferred to the material. This is known as elastic scattering, and forms the basis of contrast in a transmission electron microscope. Alternatively, the incident particle can transfer energy to the sample material, giving rise to secondary processes. For example, a secondary electron or ion can be ejected. By measuring the charge and mass of the ejected particle, useful chemical and physical information about the sample is obtained.

Three generic types of electron microscopes exist. These are shown schematically in Figure 1.5. A conventional transmission electron microscope (TEM) is shown in (a). A transparent specimen S is illuminated from above, where the illumination is omitted for simplicity. Some electrons are elastically scattered at the object point, and some remain unscattered. The unscattered current passes through an aperture A, and is imaged by a lens L onto the recording plane P, where P typically consists of an array of charged coupled devices. Some fraction of the scattered current is stopped by the aperture A. Areas of the specimen which scatter strongly thus appear dark in the image, and areas which scatter weakly appear bright. The object point is depicted as being off the cen-

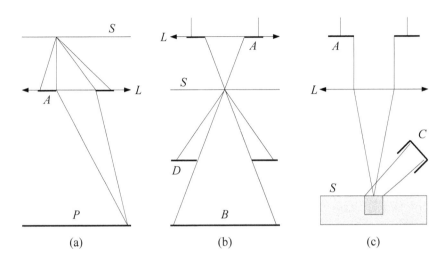

Figure 1.5: Types of electron microscopes, schematic.

tral axis. Actually, all object points in the specimen S are imaged simultaneously.

A scanning transmission electron microscope (STEM) is shown in (b). An aperture A is illuminated from above. The transmitted current is focused by a lens L onto a transparent specimen S, and is scanned sequentially over the specimen in a raster pattern. Again, some electrons are elastically scattered by the specimen, and some remain unscattered. Some fraction of the scattered current is measured on the annular dark field detector D, and the resulting signal sent to a display which is scanned synchronously with the beam of the microscope. Areas of the specimen which scatter strongly thus appear bright on the display, and areas which scatter weakly appear dark. This is not an image in the sense of Figure 1.4, because the intensity of each pixel is determined sequentially. However, it does produce an intensity map of the specimen which is just as useful as an optically formed image. The current which passes through the annular dark field detector D is measured on a bright field detector B. This signal can alternatively be displayed, with strongly scattering regions appearing dark, and weakly scattering regions appearing bright. The ultimate resolution of the TEM and STEM is the same, but the contrast differs in the two cases, depending on the accelerating voltage and numerical aperture chosen. In both cases, the numerical aperture is equal to the beam semi-angle subtended by the aperture A at the specimen S.

A scanning electron microscope (SEM) is shown in (c). An aperture A is illuminated from above. The transmitted current is focused by a lens L onto an opaque bulk specimen S, and is scanned sequentially over the specimen in a raster pattern. Low energy secondary electrons are excited by the beam in the interaction volume depicted by the darker area of the specimen S. These secondary electrons are accelerated to a collector C, and the current thus detected is used to form a signal which is sent to the display.

The ultimate resolution of the SEM is roughly equal to the size of the interaction volume, which is typically on the order of a few nanometers. One nanometer is one-billionth of a meter, and will be abbreviated $1\,nm$ throughout the text. This is a very good resolution, compared with a typical light microscope, for which the resolution is typically a few hundred nm. In addition, an SEM has superior depth of focus. This means that one need not focus precisely, in order to obtain a sharp image, allowing a seemingly three-dimensional depiction of a bulk sample. This is shown in Figure 1.6, courtesy of L.T. Varghese and L. Fan, Purdue University [90].

The schematic depiction in Figure 1.5(c) applies equally well to a scanning ion microscope (SIM), where the beam consists of ions rather than electrons. A bright source of helium ions can be formed using a sharp tip in a low pressure helium gas. The tip is elevated to a potential of a few tens of kilovolts relative to the surrounding chamber, causing a high electric field around the tip. Helium gas atoms are polarized in the field gradient, and attracted to the tip, where they dissociate to form positive helium ions. The ions are accelerated away from the tip by the electric field to form the ion beam.

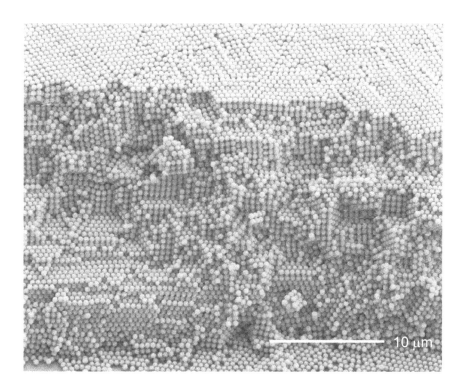

Figure 1.6: SEM picture of self-assembled silica spheres, showing high depth of focus.

A scanning helium ion micrograph is shown in Figure 1.7, obtained using an Orion SIM available commercially from Carl Zeiss SMT. A scannng electron micrograph of the same specimen is shown for comparison, obtained using a Leo SEM also available commercially from Carl Zeiss SMT. The full-scale vertical dimension is $6\,\mu$m, and the beam energy is 20 KeV in both cases. The helium ion micrograph shows striking surface detail. This is due to the fact that the helium ions are stopped within a few tens of nanometers of the surface, while the electrons penetrate several microns into the material. As a result, the material appears more transparent to electrons than to helium ions. The electron micrograph shows more contrast due to the different materials present. This is due to the fact that materials with high atomic number and high mass density preferentially scatter the electrons much more strongly than low atomic number and low mass density materials.

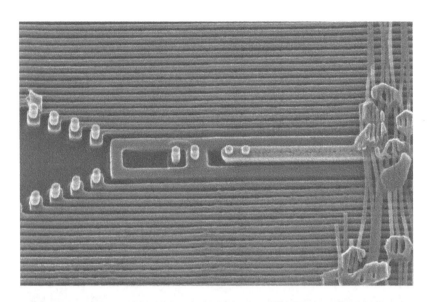

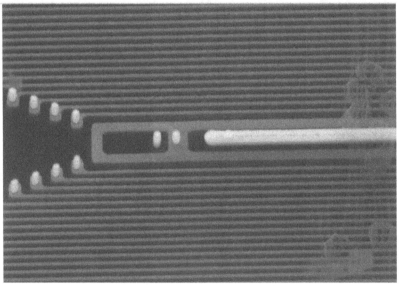

Figure 1.7: (Top) Scanning helium ion micrograph, (bottom) scanning electron micrograph.

This gives rise to high material contrast in the scanning electron microscope.

The ultimate resolution of the TEM and STEM is limited by spherical aberration and diffraction. The spherical aberration can be substantially corrected in a modern TEM and STEM, making both instruments capable of resolution in the range of 0.05 nm. This is more than sufficient to form an image of a single atom. An example of a corrected STEM image is shown in Figure 1.8. The specimen is graphene, which consists of one or more atomic

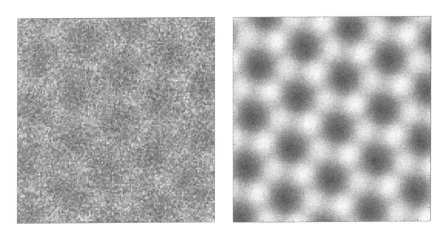

Figure 1.8: Aberration-corrected STEM images of graphene.

layers of graphite. A single layer of graphene is one atomic layer of carbon in its hexagonal crystalline form. The image on the left is a single scan recorded at 60 KV accelerating voltage in a Nion aberration-corrected STEM. The bright spots are single carbon atoms with nearest-neighbor spacing of 0.14 nm. The image on the right is derived by digitally superimposing 350 different areas of the larger image, with each area consisting of 128×128 pixels. This averages out the noise in the individual scans, without having to resort to smoothing algorithms. (The individual pixels are visible in the two images). This annular dark field image is remarkable in several respects. First, single atoms of carbon are clearly resolved with resolution better than 0.1 nm. Second, the atomic number of carbon is six, which is low relative to most solid materials. The specimen is therefore weakly scattering everywhere, thus limiting the available contrast. The fact that the contrast is adequate is remarkable. Third, the fragile graphene structure is undamaged by the beam. It would not be possible to obtain such an image without spherical aberration correction.

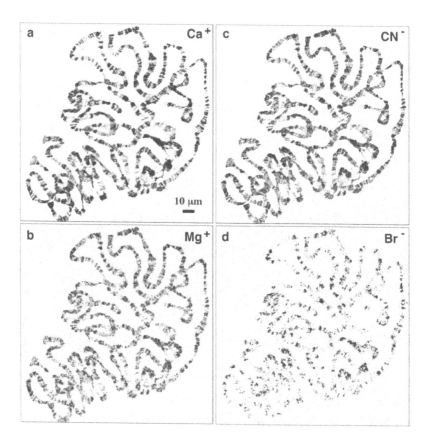

Figure 1.9: SIMS images of chromosomes.

Alternatively, a beam of ions can be used to perform chemical analysis of a material. The ion beam is focused and scanned over the surface of the material to be analyzed. Atoms are removed from the surface and ionized. These secondary ions then pass through a spectrometer which separates the various ionic species according to their masses. An image is formed synchronously, consisting of any chosen individual chemical species. This is called *Secondary Ion Mass Spectrometry* or SIMS. An example is shown in Figure 1.9

[58]. The sample consists of chromosomes of the fruit fly *Drosophila melanogaster*. Four different chemical species are shown, giving detailed spatially resolved chemical information about the chromosomes. These images were obtained using the SIMS tool at the University of Chicago, which uses a focused primary ion beam consisting of gallium ions from a liquid metal ion source. This method can be used to analyze an enormous variety of samples at the microscopic level.

Alternatively, an electron or ion beam can physically or chemically alter the target material locally. The writing substrate is coated with a thin film of organic material. Bombarding the film with a focused electron or ion beam renders the film either more soluble (positive-tone process) or less soluble (negative-tone process) in the developer. The organic film is thus patterned, and forms a binary mask for subsequent process steps. Creating fine patterns on a substrate is commonly referred to as *lithography*. An enormous variety of useful devices can be fabricated with high areal density and very small feature sizes.

Two patterns written by electron beam lithography are shown in Figure 1.10. The top pattern shows the negative-tone resist which is left behind after the development step. It is an electronic circuit pattern with 30 nm features, courtesy of Vistec Lithography. The bottom pattern shows pillars of silicon which are 0.5 μm in diameter and 1.5 μm high. They were written using a Vistec SB352 HR electron beam system, courtesy of IMS CHIPS, Stuttgart, Germany.

A focused electron beam is the smallest, finest practical writing pencil known. An arbitrary pattern can be created and stored using standardized computer-aided design software, and subsequently transmitted to the electron beam writer for one or more exposures. This flexibility, together with the high resolution, make electron beam lithography the method of choice for creating patterns on the nanometer scale of dimensions in low volume.

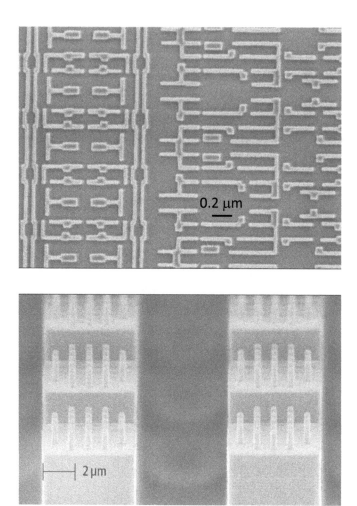

Figure 1.10: Electron beam lithography patterns.

In summary, the inherent high resolution, together with the unique interactivity with matter thus constitute two fundamental advantages of charged particle beams. They make charged particle beams indispensible to science and technology on the nanometer scale of dimensions. With this introduction, we are now in a position to begin our analytical study in detail.

Chapter 2

Geometrical optics

Geometrical optics of charged particle beams begins with relativistic classical mechanics, specifically, the motion of a charged particle in the presence of external electric and magnetic fields. The fields exert an instantaneous resultant force on the particle, which determines the path of motion. Mathematically, the solution consists of finding the three-vectors for position $\mathbf{x}$ and velocity $\mathbf{v}$ at any time t, given initial values at time zero, taking account of the influence of the fields.

Having found a prescription for solving a general particle trajectory, we can then apply this to families of trajectories. This permits us to delineate the geometrical optical properties of a beam of particles. We begin with a review of relativistic classical mechanics, focusing only on those specific topics which will lead directly to geometrical optics.

2.1 Relativistic classical mechanics

In classical mechanics, a system is described by one or more generalized coordinates Q_j, where

$$\{Q_j\} = Q_1, Q_2, \ldots, Q_n, \tag{2.1}$$

and n is the number of degrees of freedom needed to completely specify the system. For example, a three-dimensional Cartesian coordinate system can be used to completely specify position in ordinary space, and has three degrees of freedom.

The Q_j evolve under the influence of forces, and therefore depend implicitly on the time t. There exist velocities $\dot{Q}_j$ given by

$$\{\dot{Q}_j\} = \dot{Q}_1, \dot{Q}_2, \ldots, \dot{Q}_n, \tag{2.2}$$

where the dot denotes differentiation with respect to time, i.e.,

$$\dot{Q}_j \equiv \frac{d}{dt} Q_j. \tag{2.3}$$

This is quite general, since n can take on any positive integral value. For example, a system of N interacting particles has $n = 3N$ degrees of freedom.

The central problem of classical mechanics can be stated as follows: given a set of coordinates Q_j and velocities $\dot{Q}_j$ at an initial time t_0, calculate the Q_j and $\dot{Q}_j$ at any time t. The result of this calculation represents a complete specification of the system. In the present study, we will confine our attention to a single particle of rest mass m and charge q under the influence of electric and magnetic forces. We therefore define generalized coordinates $x_j = (x_1, x_2, x_3)$ with corresponding velocities $v_j = (v_1, v_2, v_3)$, where the six-vector components are functions of time t. In this case the central problem is to calculate these quantities. The prescription is general with respect to the choice of coordinate systems. For example, one could use Cartesian, cylindrical, spherical, or other coordinates with three degrees of freedom to specify the position. The reader is referred to the book by Goldstein [35] for a thorough and detailed discussion of classical mechanics.

2.1.1 Hamilton's principle of least action

We seek a general condition governing the motion of a particle with charge q and rest mass m in external electric and magnetic fields. We require that this condition be covariant with respect to the Lorentz transformation of special relativity. This ensures that the equations of motion have the same form in all frames of reference in uniform motion with respect to one another. To this end, following Goldstein, et. al. [35], we define a function $\mathcal{L}$, called the invariant Lagrangian, as

$$\mathcal{L} = \sum_{\mu=1}^{4} (m\, U_\mu\, U_\mu + q\, A_\mu\, U_\mu). \tag{2.4}$$

Here U_μ and A_μ are the four-vector velocity and electromagnetic potential, respectively, given by

$$\begin{aligned} U_\mu &= (\gamma\, \mathbf{v},\, i\gamma c) \\ A_\mu &= (\mathbf{A},\, i\phi/c), \end{aligned} \tag{2.5}$$

where $\mathbf{v}$ is the three-vector particle velocity, $\mathbf{A}$ is the magnetic three-vector potential, and ϕ is the electrostatic scalar potential. We have defined a quantity γ as

$$\gamma = \frac{1}{\sqrt{1 - v^2/c^2}}. \tag{2.6}$$

We notice from the form of (2.4) that the invariant Lagrangian $\mathcal{L}$ is a sum of inner products of two four-vectors. It is straightforward to show that the inner product of two four-vectors is invariant under a Lorentz transformation. It follows that $\mathcal{L}$ is Lorentz invariant, and has the same value in every uniformly moving reference frame. The proof of this is left to the reader in the problems.

All relevant information about the magnetic and electric fields
is contained in the magnetic vector potential $\mathbf{A}(\mathbf{x}, t)$ and electrostatic scalar potential $\phi(\mathbf{x}, t)$, respectively. In general they are
functions of position $\mathbf{x}$ and time t, measured in the particular reference frame of interest. The potentials arise from source currents
and charges which are distributed in proximity to the charged particle of interest. They also include the effects of magnetic materials
and dielectrics. We assume in the following analysis that the po-
tentials $\mathbf{A}(\mathbf{x}, t)$ and $\phi(\mathbf{x}, t)$ are known. The reader is referred to a
definitive text by Jackson [48], which describes how to calculate
these potentials, given a known distribution of charges, currents,
conductors, dielectrics, and magnetic materials.

At this point we form a key postulate, namely, for physically allowable motion of the particle, the integral of $\mathcal{L}$ over time is stationary
with respect to first-order variation as follows:

$$\delta \int_{\tau_a}^{\tau_b} \mathcal{L} \, d\tau = 0, \tag{2.7}$$

where τ is the time measured in the rest frame of the particle, commonly known as the *proper* time. We assume that the end times τ_a
and τ_b remain fixed with respect to the variation. This expression
is also Lorentz invariant, because it is constructed wholly from
Lorentz-invariant quantities.

It is possible to construct a general covariant theory which describes the motion in every reference frame. However, for our purpose here we are interested in the particle motion in a single reference frame which is at rest relative to the laboratory, commonly
known as the *lab* frame. It greatly simplifies the discussion if we
confine our attention to this single frame. In the lab frame we can
express (2.7) in the equivalent form

$$\delta \int_{t_a}^{t_b} L \, dt = 0, \tag{2.8}$$

where we have defined $L = \mathcal{L}/\gamma$ and $t = \gamma\tau$ as the Lagrangian
and time, respectively, expressed in the lab frame. The time t is

related to the proper time τ by a Lorentz transformation, where we assume the particle coordinate is zero in the particle rest frame. Substituting (2.5) into (2.4), it follows that

$$L(\mathbf{x}, \mathbf{v}; t) = -m c^2 \sqrt{1 - v^2/c^2} + q\,\mathbf{v} \cdot \mathbf{A}(\mathbf{x}, t) - q\,\phi(\mathbf{x}, t) \quad (2.9)$$

in the lab frame. We have made use of the vector notation $\mathbf{v} \cdot \mathbf{A}$ to express the inner product of the two three-vectors $\mathbf{v}$ and $\mathbf{A}$. In Cartesian coordinates this is $\mathbf{v} \cdot \mathbf{A} = v_x A_x + v_y A_y + v_z A_z$.

The Lagrangian L is a scalar function of the position $\mathbf{x}$, and the velocity $\mathbf{v}$. The time t is regarded as a parameter which uniquely specifies a point along the particle trajectory. The position and velocity depend implicitly on the time. Indeed, the central problem is to solve for this dependence. In the case where the electric and magnetic fields vary with time, the electromagnetic potentials have explicit time dependence. For static fields, these potentials have no explicit time dependence. The Lagrangian therefore has no explicit time dependence in this case.

The integral in (2.8) can be abbreviated as

$$S_{ab} = \int_{t_a}^{t_b} L(\mathbf{x}, \mathbf{v}; t)\, dt. \quad (2.10)$$

It is a scalar quantity with units of energy times time, or *action*. The integral S_{ab} is therefore known as the action integral. The expression (2.8) says that the action integral has an extremum for the physically allowable trajectory. This trajectory exists among many hypothetical trajectories, each displaced infinitesimally from the physical trajectory. The expression (2.8) is known as *Hamilton's principle of least action*.

Forming a Taylor expansion of the variation (2.7) in the lab frame, and retaining only terms to first order in δx_i and δv_i, we find

$$\delta \int_{t_a}^{t_b} L\, dt = \int_{t_a}^{t_b} \sum_{i=1}^{3} \left(\frac{\partial L}{\partial x_i} \delta x_i + \frac{\partial L}{\partial v_i} \delta v_i \right) dt. \quad (2.11)$$

Making use of the chain rule for partial derivatives, we have

$$\frac{d}{dt}\left(\frac{\partial L}{\partial v_i}\delta x_i\right) = \delta x_i \frac{d}{dt}\left(\frac{\partial L}{\partial v_i}\right) + \frac{\partial L}{\partial v_i}\delta v_i, \qquad (2.12)$$

where $\delta v_i = (d/dt)\delta x_i$. It follows (2.11, 2.12) that

$$\delta \int_{t_a}^{t_b} L\, dt = \sum_{i=1}^{3}\left[\frac{\partial L}{\partial v_i}\delta x_i\right]_{t_a}^{t_b} + \sum_{i=1}^{3}\int_{t_a}^{t_b}\delta x_i \left(\frac{\partial L}{\partial x_i} - \frac{d}{dt}\frac{\partial L}{\partial v_i}\right) dt = 0. \qquad (2.13)$$

Now $\delta t_a = \delta t_b = 0$, because the end times t_a and t_b are assumed to be fixed. This in turn demands $\delta x_i = v_i\,\delta t = 0$ at the end times t_a and t_b. Consequently, the first term on the right of (2.13) is zero. Since δx_i inside the integral is arbitrary, it is a necessary condition that

$$\frac{\partial L}{\partial x_i} - \frac{d}{dt}\frac{\partial L}{\partial v_i} = 0, \qquad (2.14)$$

where $i = 1, \ldots, 3$. This is a set of three coupled equations, known as the Euler-Lagrange equations of motion. Given the Lagrangian (2.9) and the initial conditions for position x_i and velocity v_i at time zero, these equations can be solved in principle for the components x_i and v_i as functions of time. This represents a solution to the central dynamical problem for a single particle. We will investigate the solution in more detail in the coming sections.

It is straightforward to show (2.9, 2.14) that

$$\frac{d}{dt}\left(\gamma\, m\, \mathbf{v}\right) = q\left(\mathbf{E} + \mathbf{v} \times \mathbf{B}\right) \qquad (2.15)$$

where we have defined the three-vector electric and magnetic fields, respectively, as

$$\begin{aligned} \mathbf{E} &= -\nabla\phi - \frac{\partial \mathbf{A}}{\partial t}, \\ \mathbf{B} &= \nabla \times \mathbf{A}, \end{aligned} \qquad (2.16)$$

and we have made use of the total time derivative

$$\frac{d}{dt} = \frac{\partial}{\partial t} + \mathbf{v} \cdot \nabla. \qquad (2.17)$$

The proof of (2.15) is left to the reader in the problems.

We define the three-vector kinetic momentum **p** as

$$\mathbf{p} = \gamma\, m\, \mathbf{v}. \tag{2.18}$$

Equation (2.15) is an expression of Newton's law of motion for a charged particle, where the left side is the time rate of change of the kinetic momentum, and the right side is known as the Lorentz force.

In principle it is possible to calculate all of particle optics by solving (2.15), for the position **x** and the velocity **v** as functions of time t, but further considerations will lead to a more detailed understanding, and to greater computational efficiency.

Problems

1. An arbitrary four-vector $A_\mu = (A_1, A_2, A_3, A_4)$ is defined in terms of its four components. For two reference frames in relative uniform motion with velocity v along the z-direction, the components of A_μ are related in the two frames by

$$
\begin{aligned}
A_1' &= A_1 \\
A_2' &= A_2 \\
A_3' &= \gamma\,(A_3 + i\beta A_4) \\
A_4' &= \gamma\,(-i\beta A_3 + A_4),
\end{aligned}
$$

where γ is given by (2.9) and $\beta = v/c$. This is known as a Lorentz transformation. Show that the inner product of any two four-vectors A_μ and B_μ satisfies

$$\sum_{\mu=1}^{4} A_\mu' B_\mu' = \sum_{\mu=1}^{4} A_\mu B_\mu.$$

An inner product of two four-vectors is thus said to be invariant under a Lorentz transformation.

2. Derive the Lorentz force law (2.15) from the Euler-Lagrange equations of motion (2.14).

2.1.2 The Hamiltonian function and energy conservation

We *define* a new function H by

$$H(\mathbf{x}, \mathbf{P}; t) = \sum_{i=1}^{3} P_i v_i - L(\mathbf{x}, \mathbf{v}; t), \tag{2.19}$$

where $\mathbf{P}$ is an arbitrary three-vector, whose meaning will become clear in the following. The scalar function H is derived from the Lagrangian L by a specific transformation called a *Legendre transformation* [72]. We form the total time derivative of H by invoking the chain rule,

$$\frac{dH}{dt} = \sum_{i=1}^{3} \left(\frac{\partial H}{\partial x_i} \frac{dx_i}{dt} + \frac{\partial H}{\partial P_i} \frac{dP_i}{dt} \right) + \frac{\partial H}{\partial t}, \tag{2.20}$$

recalling that $v_i = dx_i/dt$. From the definition (2.19) we obtain the identities

$$\frac{\partial H}{\partial x_i} = -\frac{\partial L}{\partial x_i}, \qquad \frac{\partial H}{\partial P_i} = v_i, \qquad \frac{\partial L}{\partial v_i} = P_i, \qquad \frac{\partial H}{\partial t} = -\frac{\partial L}{\partial t}. \tag{2.21}$$

The third of these, together with (2.14) leads to

$$\frac{\partial L}{\partial x_i} = \frac{dP_i}{dt}. \tag{2.22}$$

It follows that the large parenthesis in (2.20) vanishes identically, and

$$\frac{dH}{dt} = -\frac{\partial L}{\partial t}. \tag{2.23}$$

The function H is called the *Hamiltonian* functon, and the three-vector $\mathbf{P}$ is called the *canonical momentum*. From (2.9) and the third identity (2.21), the canonical momentum components can be written as

$$P_i = \gamma m v_i + q A_i, \qquad i = 1, \ldots, 3. \qquad (2.24)$$

Equivalently, from (2.18), this is

$$P_i = p_i + q\, A_i. \qquad (2.25)$$

The canonical momentum is thus the sum of the kinetic momentum plus the charge times the magnetic vector potential. Obviously, the canonical momentum and kinetic momentum are identical in the case where the magnetic vector potential is zero.

Next we consider the special case where the potentials $\mathbf{A}$ and ϕ have no explicit time dependence; i.e., the fields are static. From (2.9) it follows that the right side of (2.23) vanishes, and

$$\frac{dH}{dt} = 0. \qquad (2.26)$$

This means that H is a conserved quantity in this case. From (2.9, 2.19, 2.24, 2.26) it follows that

$$H = \gamma m c^2 + q\phi = const, \qquad (2.27)$$

and H is a constant of the motion. We will see in the following that H can be identified with the total energy.

The energy H does not depend on the magnetic vector potential $\mathbf{A}$, because the magnetic Lorentz force in (2.15) acts in a direction perpendicular to the particle velocity $\mathbf{v}$. As a result, the magnetic force alters the direction of the velocity $\mathbf{v}$, but not the magnitude. Consequently, the magnetic force cannot cause a change in energy.

We now proceed to define two quantities which will prove very useful later. We define a quantity E by

$$E = \gamma m c^2, \qquad (2.28)$$

where mc^2 is the rest energy. We further define the kinetic energy T by

$$E = T + mc^2. \qquad (2.29)$$

By this definition, the energy E is the sum of the kinetic energy plus the rest energy. The Hamiltonian H is then

$$H = T + mc^2 + q\phi. \tag{2.30}$$

The Hamiltonian is the sum of the kinetic energy plus the rest energy plus the potential energy. The Hamiltonian H thus represents the total energy. It is conserved in the case where the potentials ϕ and $\mathbf{A}$ have no explicit time dependence. Any force which acts in such a way that the total energy is constant is called a *conservative* force.

Problems

1. Show from the above analysis that

$$E^2 = p^2 c^2 + m^2 c^4, \tag{2.31}$$

where $p^2 \equiv \mathbf{p} \cdot \mathbf{p}$.

2. Prove the identity

$$pc = \beta E, \tag{2.32}$$

where $\beta = v/c$.

2.1.3   Mechanical analog of Fermat's principle

We now concentrate on the important special case where the electric and magnetic fields are constant in time. Mathematically, this is equivalent to the potentials $\mathbf{A}(\mathbf{x}, t) \equiv \mathbf{A}(\mathbf{x})$ and $\phi(\mathbf{x}, t) \equiv \phi(\mathbf{x})$ having no explicit time dependence. We showed in the preceding section that the Hamiltonian represents the conserved total energy in this case (2.27). We now define a quantity W_{ab} as the component of the canonical momentum $\mathbf{P}$ along the trajectory path, integrated over the path between the two endpoints $\mathbf{x}_a$ and $\mathbf{x}_b$. It is given by

$$W_{ab} = \int_{\mathbf{x}_a}^{\mathbf{x}_b} \mathbf{P} \cdot d\mathbf{s}, \tag{2.33}$$

where the integration path is assumed to be the path of physically allowable particle motion, satisfying (2.14, 2.27). Equivalently,

$$W_{ab} = \int_{t_a}^{t_b} \left(\sum_{i=1}^{3} P_i \, v_i \right) dt. \tag{2.34}$$

The function W_{ab} is the integral of the *action* along the path. It is also known as the *eikonal* function. From (2.19),

$$W_{ab} = \int_{t_a}^{t_b} (L + H) \, dt = \int_{t_a}^{t_b} L \, dt + H \, (t_b - t_a), \tag{2.35}$$

where, in the rightmost equality, we only consider possible motion for which $H = const$. The variation is

$$\delta W_{ab} = \delta \int_{t_a}^{t_b} L \, dt + H \, (\delta t_b - \delta t_a). \tag{2.36}$$

This variation is shown schematically in Figure 2.1, where the solid curve represents the physically allowable path, and the broken curve represents an infinitesimally displaced path, which is *not* physically allowable. The endpoints are held fixed by assumption in the variation. In order that $H = const$, it is necessary to allow the end times t_a and t_b to vary. This is different from Hamilton's principle (2.7), where the end times t_a and t_b are assumed to be fixed. Consequently, in the present case,

$$
\begin{aligned}
\delta \int_{t_a}^{t_b} L \, dt \;=\;& \left(\delta t_a \frac{\partial}{\partial t_a} + \delta t_b \frac{\partial}{\partial t_b} \right) \int_{t_a}^{t_b} L \, dt \\
&+ \int_{t_a}^{t_b} \sum_{i=1}^{3} \left(\frac{\partial L}{\partial x_i} \delta x_i + \frac{\partial L}{\partial v_i} \delta v_i \right) dt
\end{aligned} \tag{2.37}
$$

where the first term on the right accounts for the variation of the end times t_a and t_b, and the second term accounts for the variation

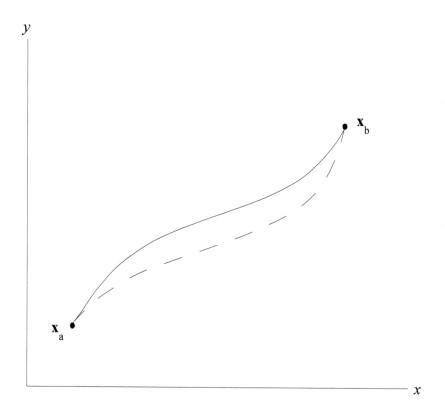

Figure 2.1: Variation of the particle trajectory for fixed endpoints

of the integrand. The integrand of the second term on the right can be rewritten (2.14) as

$$\frac{\partial L}{\partial x_i} \delta x_i + \frac{\partial L}{\partial v_i} \delta v_i = \delta x_i \frac{d}{dt} \left(\frac{\partial L}{\partial v_i} \right) + \frac{\partial L}{\partial v_i} \frac{d}{dt} (\delta x_i) = \frac{d}{dt} \left(\frac{\partial L}{\partial v_i} \delta x_i \right).$$
(2.38)

From the third identity of (2.21) together with (2.37, 2.38),

$$\delta \int_{t_a}^{t_b} L \, dt = \left[L \, \delta t \right]_{t_a}^{t_b} + \sum_{i=1}^{3} \left[P_i \, \delta x_i \right]_{t_a}^{t_b}.$$
(2.39)

We now impose the condition that the endpoints $\mathbf{x}_a$ and $\mathbf{x}_b$ remain fixed. To ensure this, we require that $\delta x_i = -v_i \, \delta t$ at the end times t_a and t_b, to compensate for what would otherwise be an offset of the endpoints. From (2.36, 2.39)

$$\delta W_{ab} = \left[\left(-\sum_{i=1}^{3} P_i \, v_i + L + H \right) \delta t \right]_{t_a}^{t_b} \equiv 0, \qquad (2.40)$$

where the large parenthesis vanishes identically from (2.19). This is the principle of least action for the special case where the potentials $\mathbf{A}(\mathbf{x})$ and $\phi(\mathbf{x})$ contain no explicit time dependence. This can also be written (2.33) as

$$\delta \int_{\mathbf{x}_a}^{\mathbf{x}_b} \mathbf{P} \cdot d\mathbf{s}, = 0 \qquad (2.41)$$

where the endpoints $\mathbf{x}_a$ and $\mathbf{x}_b$ are assumed to be fixed. The integral has units of action. The equation (2.41) can be regarded as the principle of least action for the case where the potentials have no explicit time dependence. We have shown that this is a necessary condition for physically allowable motion.

We define a scalar quantity n as the component of canonical momentum along the path of motion (2.25):

$$n = \mathbf{P} \cdot \hat{\mathbf{s}} = p + q \, \mathbf{A} \cdot \hat{\mathbf{s}}, \qquad (2.42)$$

where $\hat{\mathbf{s}}$ is the unit vector along the direction of motion, locally tangent to the trajectory, and p is the scalar kinetic momentum. From (2.41, 2.42), the principle of least action can also be written as

$$\delta \int_{\mathbf{x}_a}^{\mathbf{x}_b} n \, ds = 0. \qquad (2.43)$$

A close analogy exists with light optics. Fermat's principle states that light propagates along that path which minimizes the transit time between two points. This can be written as a variational principle as follows:

$$\delta \int_{t_a}^{t_b} dt = 0. \qquad (2.44)$$

The speed of light is path length traversed per unit time, or ds/dt, where ds is the element of path length. From the Maxwell theory, an electromagnetic wave travels with phase velocity v given by

$$v = \frac{c}{n'}, \qquad (2.45)$$

where n' is the index of refraction of the medium, and $n' = 1$ in vacuum. Substituting,

$$\delta \int_{\mathbf{x}_a}^{\mathbf{x}_b} n' \, ds = 0. \qquad (2.46)$$

The physical path taken by the light is that path for which integral in (2.46) is a minimum. The equations (2.43) and (2.46) are formally identical, expressing a close analogy between light propagation and particle propagation. The index of refraction n' for light varies in general with position within the medium. The quantity n in (2.42) is identified with an index of refraction for a particle. It depends on the electrostatic potential $\phi(\mathbf{x})$ through the momentum p, and depends on the magnetic vector potential $\mathbf{A}(\mathbf{x})$ explicitly. The electromagnetic potential varies slowly in space, as governed by Maxwell's equations.

Formulation of the dynamical problem in this way has the advantage that it does not rely on time as an explicit parameter, as long as the potentials are time independent. This greatly simplifies the discussion of geometrical optics for this important class of problems. For example, in many particle beam instruments we are only interested in where the ray ends, but not in the time at which the particle arrives.

In the following sections, we will make use of the variational principle (2.43) to solve for the detailed physical trajectory.

2.2   Exact trajectory equation for a single particle

We now make use of the preceding analysis to find an explicit differential equation governing particle motion for time independent potentials. The following analysis closely follows that of Sturrock

[86]. We seek a condition, based on the principle of least action for time independent potentials, which will allow a solution for the position **x** at all points along a physical trajectory. Expanding the variation (2.43) we have

$$\delta W_{ab} = \delta \int_{\mathbf{x}_a}^{\mathbf{x}_b} n \, ds = \int_{\mathbf{x}_a}^{\mathbf{x}_b} \left[\delta n + n \frac{d}{ds}(\delta s) \right] ds, \qquad (2.47)$$

where we assume the endpoints $\mathbf{x}_a$ and $\mathbf{x}_b$ remain fixed. The first term in the square bracket is the variation of the refractive index, and the second term is the variation of the path of integration. We assume for now that this applies to an arbitrary path, not necessarily a physically allowable trajectory.

Expanding the differential path length ds in terms of the position $d\mathbf{x}$, we find

$$(ds)^2 = d\mathbf{x} \cdot d\mathbf{x}. \qquad (2.48)$$

Taking the differential of both sides

$$(ds)\,\delta(ds) = d\mathbf{x} \cdot \delta(d\mathbf{x}). \qquad (2.49)$$

The unit vector $\hat{\mathbf{s}}$ along the path can be written as

$$\hat{\mathbf{s}} = \frac{d\mathbf{x}}{ds}. \qquad (2.50)$$

Interchanging the order of differentials in (2.49), it follows (2.50) that

$$\frac{d}{ds}(\delta s) = \hat{\mathbf{s}} \cdot \frac{d}{ds}(\delta \mathbf{x}), \qquad (2.51)$$

from which

$$\delta \hat{\mathbf{s}} = \delta \left(\frac{d\mathbf{x}}{ds} \right) = \frac{d}{ds}(\delta \mathbf{x}) - \hat{\mathbf{s}} \left[\hat{\mathbf{s}} \cdot \frac{d}{ds}(\delta \mathbf{x}) \right]. \qquad (2.52)$$

Expanding the variation δn,

$$\delta n = \nabla_{\mathbf{x}} n \cdot \delta \mathbf{x} + \nabla_{\hat{\mathbf{s}}} n \cdot \delta \hat{\mathbf{s}}, \qquad (2.53)$$

where (2.42)

$$\nabla_{\hat{\mathbf{s}}} n = q \, \mathbf{A}. \qquad (2.54)$$

We obtain an expression (2.42, 2.52, 2.53, 2.54) for the integrand in (2.48) as

$$\delta n + n \frac{d}{ds}(\delta s) = (\nabla_{\mathbf{x}} n) \cdot \delta \mathbf{x} + \mathbf{P} \cdot \frac{d}{ds}(\delta \mathbf{x}). \tag{2.55}$$

The chain rule gives

$$\frac{d}{ds}(\mathbf{P} \cdot \delta \mathbf{x}) = \mathbf{P} \cdot \frac{d}{ds}(\delta \mathbf{x}) + \frac{d\mathbf{P}}{ds} \cdot (\delta \mathbf{x}), \tag{2.56}$$

from which it follows (2.47, 2.55, 2.56)

$$\delta W_{ab} = \left[\mathbf{P} \cdot \delta \mathbf{x} \right]_{\mathbf{x}_a}^{\mathbf{x}_b} + \int_{\mathbf{x}_a}^{\mathbf{x}_b} \delta \mathbf{x} \cdot \left(\nabla_{\mathbf{x}} n - \frac{d\mathbf{P}}{ds} \right) ds. \tag{2.57}$$

We now invoke the principle of least action (2.43), namely, $\delta W_{ab} = 0$. The first term on the right is zero, as the endpoints are assumed to be fixed, i.e., $\delta \mathbf{x}_a = \delta \mathbf{x}_b = 0$. As $\delta \mathbf{x}$ under the integral on the right is arbitrary, it becomes a necessary condition that the large parenthesis in (2.57) must vanish, i.e.,

$$\nabla_{\mathbf{x}} n - \frac{d\mathbf{P}}{ds} = 0. \tag{2.58}$$

This represents the exact trajectory equation, relativistically correct in the lab frame, where we recall (2.18, 2.25, 2.42). For specified endpoints $\mathbf{x}_a$ and $\mathbf{x}_b$, this equation can be solved in principle to find the position $\mathbf{x}$ everywhere along a single trajectory of a single particle.

2.3   Conservation laws

We showed previously that, in the case where the potentials $\mathbf{A}(\mathbf{x})$ and $\phi(\mathbf{x})$ have no explicit time dependence, the total energy H is a constant of the motion. In this section, we show that other invariant quantities exist, as a direct consequence of the least action principle. As in the preceding section, the reader is referred to the book by Sturrock [86] for a detailed discussion.

2.3.1 The Lagrange invariant

In the preceding sections, we derived the necessary conditions for a single trajectory to represent physically allowable motion. Henceforth we refer to a physically allowable trajectory satisfying (2.43, 2.58) as a *ray*. In this section, we consider the behavior of rays which are infinitesimally displaced from one another. This is shown schematically in Figure 2.2. From (2.57) the variation in optical path length between two neighboring rays is given by

$$\delta W_{ab} = \mathbf{P}_b \cdot \delta \mathbf{x}_b - \mathbf{P}_a \cdot \delta \mathbf{x}_a. \tag{2.59}$$

This infinitesimal quantity is nonzero in general, since the endpoints $\mathbf{x}_a$ and $\mathbf{x}_b$ of the two rays are assumed in general to be displaced from one another. It can be shown that δW_{ab} is an *exact differential* [72], in which case

$$\mathbf{P}_b = \nabla_{\mathbf{x}_b} W_{ab}, \qquad \mathbf{P}_a = -\nabla_{\mathbf{x}_a} W_{ab}. \tag{2.60}$$

Geometrically, this means that the canonical momentum $\mathbf{P}$ is normal to surfaces of constant optical path length, $W_{ab} = const$ at the endpoints, where we note that the endpoints can be chosen to be anywhere along the ray path.

We now consider a second perturbation, independent from the first. It follows that (2.59):

$$d(\delta W_{ab}) = d\mathbf{P}_b \cdot \delta \mathbf{x}_b + \mathbf{P}_b \cdot d(\delta \mathbf{x}_b) - d\mathbf{P}_a \cdot \delta \mathbf{x}_a - \mathbf{P}_a \cdot d(\delta \mathbf{x}_a). \tag{2.61}$$

Interchanging the order of perturbations and subtracting, we obtain

$$d\mathbf{P}_a \cdot \delta \mathbf{x}_a - \delta \mathbf{P}_a \cdot d\mathbf{x}_a = d\mathbf{P}_b \cdot \delta \mathbf{x}_b - \delta \mathbf{P}_b \cdot d\mathbf{x}_b. \tag{2.62}$$

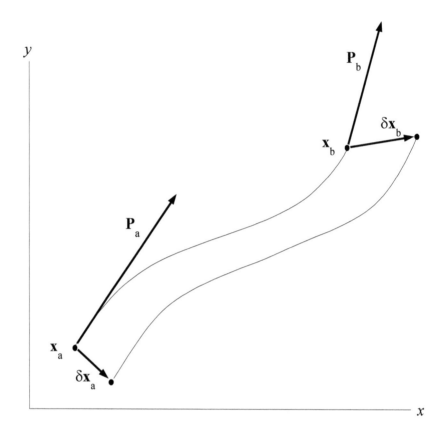

Figure 2.2: Two rays, infinitesimally displaced from one another.

Since $\mathbf{x}_a$ and $\mathbf{x}_b$ can be *any* two points connected by a ray, it follows that

$$d\mathbf{P} \cdot \delta\mathbf{x} - \delta\mathbf{P} \cdot d\mathbf{x} = const, \qquad (2.63)$$

where the δ- and d-variations refer to two separate rays, each derived by an independent perturbation from the original ray. This quantity is known as the Lagrange invariant. To appreciate the meaning of the Lagrange invariant, we consider a special case for which $d\mathbf{x}_a = \delta\mathbf{x}_b = 0$. In this case (2.62) reduces to

$$-\delta\mathbf{P}_a \cdot d\mathbf{x}_a = d\mathbf{P}_b \cdot \delta\mathbf{x}_b. \qquad (2.64)$$

This is shown schematically in Figure 2.3 , where the unperturbed

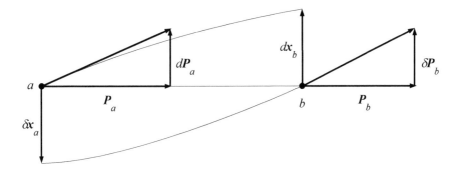

Figure 2.3: Perturbed rays for case $d\mathbf{x}_a = \delta\mathbf{x}_b = 0$.

ray is represented by a straight line connecting the beginning point a and the endpoint b. We now choose local z-axes codirectional with $\mathbf{P}_a$ and $\mathbf{P}_b$ at the respective endpoints $\mathbf{x}_a$ and $\mathbf{x}_b$. We further choose $\delta\mathbf{x}_a$ colinear (either codirectional or antidirectional) with $d\mathbf{P}_a$ and $d\mathbf{x}_b$ colinear with $\delta\mathbf{P}_b$, in the respective transverse end planes. Since $d\mathbf{P}_a$ is perpendicular to $\mathbf{P}_a$, this represents a change in direction, but not magnitude of $\mathbf{P}_a$. A similar statement holds for $\mathbf{P}_b$. The Lagrange invariant (2.62) reduces to

$$-\delta P_a\, dx_a = dP_b\, \delta x_b. \tag{2.65}$$

We notice (2.25) that $\delta P_a = p_a\, \delta\theta_a$ and $dP_b = p_b\, d\theta_b$, since the magnetic vector potential $\mathbf{A}$ is assumed to be unchanged in the perturbation. Recalling that p is the scalar kinetic momentum, it follows that

$$-p_a\, \delta\theta_a\, dx_a = p_b\, d\theta_b\, \delta x_b, \tag{2.66}$$

where dx_b is proportional to $d\theta_a$, and $\delta\theta_b$ is proportional to δx_a. Repeating this variation process in the orthogonal transverse axis, and multiplying,

$$p_a^2\, \delta\Omega_a\, dA_a = p_b^2\, d\Omega_b\, \delta A_b, \tag{2.67}$$

where $d\Omega = d\theta_x\, d\theta_y$ is the solid angle element, and $dA = dx\, dy$ is the transverse area element.

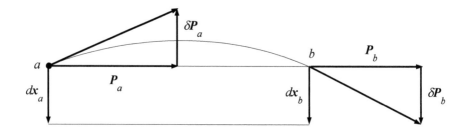

Figure 2.4: Perturbed rays for case $\delta\mathbf{x}_a = \delta\mathbf{x}_b = 0$.

Next we consider the special case where $\delta\mathbf{x}_a = \delta\mathbf{x}_b = 0$. This is shown schematically in Figure 2.4, where, again, the unperturbed ray is represented by a straight line connecting the beginning point a and the endpoint b. The two neighboring rays emanating from a single point, with infinitesimally differing directions intersect the same endpoint. In this case the endpoints $\mathbf{x}_a$ and $\mathbf{x}_b$ are said to be *optically conjugate*. Because $\delta\mathbf{x}_a = \delta\mathbf{x}_b = 0$, it follows directly from (2.59) that $\delta W_{ab} = 0$. This means that the two rays have identical optical path length W_{ab}. This is equivalent to the statement that δW_{ab} is a perfect differential, since the line integral of W_{ab} around the closed path of the two rays is zero.

The Lagrange invariant (2.62) reduces in this case to

$$d\mathbf{P}_a \cdot \delta\mathbf{x}_a = d\mathbf{P}_b \cdot \delta\mathbf{x}_b. \tag{2.68}$$

Applying the preceding method,

$$p_a \, d\theta_a \, \delta x_a = p_b \, d\theta_b \, \delta x_b. \tag{2.69}$$

This is known as the law of Helmholtz–Lagrange [86]. We define the magnification $M = \delta x_b/\delta x_a$, in which case the angular magnification is given by $d\theta_b/d\theta_a = p_a/(Mp_b)$. Repeating this in the orthogonal axis as before, it follows that

$$p_a^2 \, d\Omega_a \, \delta A_a = p_b^2 \, d\Omega_b \, \delta A_b. \tag{2.70}$$

The product of transverse area element times solid angle element is called the *emittance*. Equation (2.70) shows that the product of the emittance times the square of the momentum is conserved. For a ray bundle, the current divided by the emittance is called the *brightness*. It follows from (2.67, 2.70) that the ratio of the brightness β divided by the square of the momentum is conserved, assuming constant current. This can be written as

$$\frac{\beta}{p^2} = const, \qquad (2.71)$$

where p is the relativistic scalar kinetic momentum. It does not require that the two end planes be optically conjugate, as it applies to both of the above special cases. It follows that it is impossible to focus any beam to a spot which is brighter than the source. These arguments apply strictly only over infinitesimal regions. It is common practice to apply brightness conservation to a finite region, such as a whole beam. This is only approximate, however, and becomes less accurate as the whole beam becomes larger.

Next it is interesting to consider the special case where $\mathbf{P}_a$ is inclined by an angle θ to local z-axis. The above case becomes

$$p_a \cos\theta_a \, d\theta_a = M \, p_b \cos\theta_b \, d\theta_b. \qquad (2.72)$$

To this point, we have considered only infinitesimal perturbations of first order. It is interesting to consider the case in which rays inclined at *finite* angle θ intersect the same image point $\mathbf{x}_b$ for all θ. This corresponds to perfect imaging, without aberration. We integrate as follows:

$$p_a \int_0^{\theta_a} \cos\theta_a \, d\theta_a = M \, p_b \int_0^{\theta_b} \cos\theta_b \, d\theta_b,$$
$$p_a \sin\theta_a = M \, p_b \sin\theta_b, \qquad (2.73)$$

for all θ_a and θ_b. This is presumed true independent of $\delta\mathbf{x}_a$, in which case it represents perfect imaging with regard to all aberrations which are linear in $\mathbf{x}_a$; i.e., coma. This is known as the Abbe–Helmholtz sine condition for coma-free imaging [86].

An analogous case exists where we assume $\delta\mathbf{x}_a$ to be parallel with the axis, and $\mathbf{P}_a$ inclined at angle θ_a. We find that

$$p_a \sin\theta_a \, d\theta_a = M_L \, p_b \sin\theta_b \, d\theta_b, \tag{2.74}$$

where $M_L = \delta z_b / \delta z_a$ is defined as the longitudinal magnification. Assuming perfect imaging as before, it follows that

$$
\begin{aligned}
p_a \int_0^{\theta_a} \sin\theta_a \, d\theta_a &= M_L \, p_b \int_0^{\theta_b} \sin\theta_b \, d\theta_b, \\
p_a \sin^2(\theta_a/2) &= M_L \, p_b \sin^2(\theta_b/2), \tag{2.75}
\end{aligned}
$$

for all θ_a and θ_b. This is known as Herschel's condition for vanishing spherical aberration [86]. It follows from the preceding that the longitudinal and transverse magnifications are related by

$$M_L = M^2 \, p_b/p_a. \tag{2.76}$$

By successive applications of the Legendre transformation, it is possible to construct other characteristic functions from $W(\mathbf{x}_a, \mathbf{x}_b)$. For example, let

$$V(\mathbf{x}_a, \mathbf{P}_b) = \mathbf{P}_b \cdot \mathbf{x}_b - W(\mathbf{x}_a, \mathbf{x}_b). \tag{2.77}$$

It follows that

$$\delta V = \mathbf{P}_a \cdot \delta\mathbf{x}_a + \mathbf{x}_b \cdot \delta\mathbf{P}_b. \tag{2.78}$$

Continuing this procedure, we define

$$X(\mathbf{P}_a, \mathbf{x}_b) = \mathbf{P}_a \cdot \mathbf{x}_a + V(\mathbf{x}_a, \mathbf{x}_b). \tag{2.79}$$

It follows that

$$\delta X = \mathbf{x}_a \cdot \delta\mathbf{P}_a + \mathbf{P}_b \cdot \delta\mathbf{x}_b. \tag{2.80}$$

Similarly we define

$$Y(\mathbf{P}_a, \mathbf{P}_b) = -\mathbf{P}_a \cdot \mathbf{x}_a + V(\mathbf{x}_a, \mathbf{P}_b). \tag{2.81}$$

It follows that

$$\delta Y = -\mathbf{x}_a \cdot \delta\mathbf{P}_a + \mathbf{x}_b \cdot \delta\mathbf{P}_b. \tag{2.82}$$

The functions V, W, X, and Y represent a way to describe the optical coupling between the space $(\mathbf{x}_a, \mathbf{P}_a)$ and the space $(\mathbf{x}_b, \mathbf{P}_b)$ in an infinitesimal region surrounding a ray.

Problem

Show that the functions V, W, X, and Y all lead to the same Lagrange invariant.

2.3.2 Liouville's theorem and brightness conservation

The motion of a particle can be considered to trace out a trajectory in a six-dimensional space, for which the coordinates are labeled by the three position components of $\mathbf{x}$ and the three canonical momentum components of $\mathbf{P}$. This is called *phase space*. The reader is referred to Goldstein et. al. [35] for background and further details.

To introduce this description, we notice (2.14, 2.19) that

$$\frac{\partial H}{\partial x_j} = -\frac{\partial L}{\partial x_j} = -\frac{d}{dt}\left(\frac{\partial L}{\partial v_j}\right) = -\frac{dP_j}{dt}, \qquad (2.83)$$

and (2.19) that

$$\frac{\partial H}{\partial P_j} = v_j = \frac{dx_j}{dt}, \qquad (2.84)$$

for $j = 1, \ldots, 3$. Summarizing, this yields a coupled set of six first-order equations as follows:

$$\frac{\partial H}{\partial x_j} = -\frac{dP_j}{dt}, \qquad\qquad \frac{\partial H}{\partial P_j} = \frac{dx_j}{dt}, \qquad (2.85)$$

where $j = 1, \ldots, 3$. These are known as Hamilton's equations of motion. Given the Hamiltonian function (2.9, 2.19) together with an initial condition $(\mathbf{x}_0, \mathbf{P}_0)$ at any single phase space point along the trajectory, Hamilton's equations can be solved in principle to find the entire phase space trajectory of a single particle.

We imagine a family of trajectories, all infinitesimally displaced from one another, with each corresponding to a slightly different initial condition. These trajectories cannot intersect in phase space, as to do so would imply that a single initial condition would give rise to multiple end conditions. As such, an analogy exists with fluid flow, where the trajectories can be described by a flux $\mathbf{j}$ and a density ρ of points in phase space. As trajectories are conserved, these quantities obey a continuity equation

$$\nabla \cdot \mathbf{j} + \frac{\partial \rho}{\partial t} = 0, \tag{2.86}$$

where

$$\mathbf{j} = \rho \mathbf{v}, \tag{2.87}$$

and $\mathbf{v}$ is the six-dimensional velocity. Expanding the six-divergence,

$$
\begin{aligned}
\nabla \cdot \mathbf{j} &= \sum_{j=1}^{3} \left[\frac{\partial}{\partial x_j}(\rho \dot{x}_j) + \frac{\partial}{\partial P_j}(\rho \dot{P}_j) \right] \\
&= \sum_{j=1}^{3} \left[\rho \left(\frac{\partial \dot{x}_j}{\partial x_j} + \frac{\partial \dot{P}_j}{\partial P_j} \right) + \left(\frac{\partial \rho}{\partial x_j} \frac{dx_j}{dt} + \frac{\partial \rho}{\partial P_j} \frac{dP_j}{dt} \right) \right],
\end{aligned}
\tag{2.88}
$$

where the dot signifies total time derivative. The first term on the right vanishes by Hamilton's equations (2.85). It follows that

$$\nabla \cdot \mathbf{j} + \frac{\partial \rho}{\partial t} = \sum_{j=1}^{3} \left(\frac{\partial \rho}{\partial x_j} \frac{dx_j}{dt} + \frac{\partial \rho}{\partial P_j} \frac{dP_j}{dt} \right) + \frac{\partial \rho}{\partial t}, \tag{2.89}$$

where we recognize the right side as the total time derivative $d\rho/dt$. From (2.86, 2.89) it follows that

$$\frac{d\rho}{dt} = 0. \tag{2.90}$$

This is called Liouville's theorem. It means that $\rho = const$, and the density of trajectory points in phase space is conserved.

Applying this to a beam, the geometry is shown schematically in Figure 2.5. We imagine particles emitted from an infinitesimal

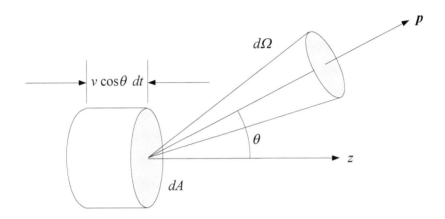

Figure 2.5: Geometry for brightness conservation.

area element dA into an infinitesimal solid angle element $d\Omega$ centered around the kinetic momentum vector $\mathbf{p}$. The phase space density is given locally by

$$\rho\left(\mathbf{x}, \mathbf{P}\right) = \frac{d^6 N}{d^3 x \, d^3 P} = const, \qquad (2.91)$$

where

$$d^3 x = v \, \cos\theta \, dt \, dA, \qquad d^3 P = p^2 \, dp \, d\Omega, \qquad (2.92)$$

and the scalar kinetic momentum p is related to the velocity v by (2.18). Passing to the limit of an infinitesimally thin volume element in the z-axis, the density ρ is a delta function in z. Integrating over all z, the result is unity, by the property of the delta function. It follows that

$$\frac{1}{p^2} \frac{dN}{dp \, dA \, d\Omega} = const. \qquad (2.93)$$

We consider the special case of a monoenergetic beam with single value p. In this case the density is a delta function in p. Integrating over all p, we obtain

$$\frac{1}{p^2} \frac{dN}{dA \, d\Omega} = const. \tag{2.94}$$

We define the brightness β as the density of trajectories per unit transverse area per unit solid angle,

$$\frac{\beta}{p^2} = const. \tag{2.95}$$

The ratio of brightness to square of the relativistic kinetic momentum is conserved. This reproduces the result (2.71) found above.

Solving (2.29) for the scalar kinetic momentum p in terms of the kinetic energy T,

$$p^2 = 2m \left(T + \frac{T^2}{2mc^2} \right) = 2meV^*, \tag{2.96}$$

where we have defined a quantity V^*, referred to by many authors as the relativistic beam voltage, in which case

$$\frac{\beta}{V^*} = const. \tag{2.97}$$

As a result of this, it follows that a beam can never be focused to a spot which is brighter than the source. This has the practical consequence that the source brightness represents a fundamentally important property of any optical system.

2.4 General curvilinear axis

For many systems, it is convenient to formulate the optics in terms of transverse coordinates in a plane which is locally perpendicular to a central optic axis. As this axis need not be a straight line, we designate it a general curvilinear axis. We designate an axial coordinate z, and transverse Cartesian coordinates $x_j = (x, y)$ for $j = (1, 2)$ in a plane locally perpendicular to the axis. We further designate ray slope components $x'_j = (x', y') = dx_j/dz$. A ray is completely specified at any plane z by its two-vector transverse position $\mathbf{x}$ and its two-vector slope $\mathbf{x}'$.

The central problem in this formulation may be stated as follows: given the transverse position $\mathbf{x}_a$ and slope $\mathbf{x}'_a$ at an arbitrary starting axial coordinate z_a, find the transverse position $\mathbf{x}_b$ and slope $\mathbf{x}'_b$ at an arbitrary ending axial coordinate z_b. This is shown schematically in Figure 2.6. It is implicit here and in the following

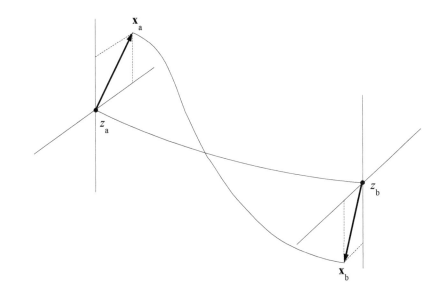

Figure 2.6: General curvilinear axis.

that the slope $\mathbf{x}'_b$ be finite. This excludes the case of a particle mirror, for which the slope is infinite where the ray turns around.

In fact, both the position and slope are considered small in the following. Equivalently, we will only investigate rays that remain close to the central axis.

Our purpose here is to identify the equations of motion, and describe a general methodology for solving them. This solution can later be applied to a large variety of specific cases, describing a similar variety of phenomena observed in practice. The reader is referred to the references by Sturrock [86], Rose [75], Hawkes and Kasper [43, 44, 45], and Wollnik [93] for further detail and elaboration. The present analysis is based on the earlier works of Glaser [33] and Sturrock [86].

2.4.1   Equation of motion in terms of transverse coordinates and slopes

We found previously that the optical path length along a ray joining two endpoints $\mathbf{x}_a$ and $\mathbf{x}_b$ is given by the action integral (2.33) as

$$W_{ab} = \int_{\mathbf{x}_a}^{\mathbf{x}_b} n \, ds = \int_{z_a}^{z_b} m \, dz, \qquad (2.98)$$

where we have defined a modified refractive index m as

$$m(\mathbf{x}, \mathbf{x}'; z) = n \frac{ds}{dz} = n \sqrt{1 + x'^2 + y'^2}, \qquad (2.99)$$

where $\mathbf{x}$ and $\mathbf{x}'$ are the two-dimensional vector position and slope components in the transverse plane, respectively, and where the prime represents differentiation with respect to z. The variation of optical path length is given by

$$\delta W_{ab} = \delta \int_{z_a}^{z_b} m \, dz = \int_{z_a}^{z_b} (\delta m) \, dz = 0. \qquad (2.100)$$

We have purposely excluded the second term in square brackets of (2.47). This is equivalent to assuming the variation of the path length along the optic axis is zero. In order for this to be meaningful, the optic axis must itself be a physical ray in the sense of satisfying (2.58).

Expanding the variation δm,

$$\delta m = \sum_{j=1}^{2} \left(\frac{\partial m}{\partial x_j} \delta x_j + \frac{\partial m}{\partial x'_j} \delta x'_j \right). \tag{2.101}$$

Using the chain rule, we find that

$$\frac{d}{dz} \left(\frac{\partial m}{\partial x'_j} \delta x_j \right) = \delta x_j \frac{d}{dz} \left(\frac{\partial m}{\partial x'_j} \right) + \frac{\partial m}{\partial x'_j} \delta x'_j. \tag{2.102}$$

This leads to

$$\delta W_{ab} = \left[\sum_{j=1}^{2} \frac{\partial m}{\partial x'_j} \delta x_j \right]_{z_a}^{z_b} + \int_{z_a}^{z_b} \sum_{j=1}^{2} \delta x_j \left(\frac{\partial m}{\partial x_j} - \frac{d}{dz} \frac{\partial m}{\partial x'_j} \right) dz = 0. \tag{2.103}$$

Assuming the endpoints are fixed, $\delta x_j = 0$ at z_a and z_b, the square bracket vanishes. Furthermore, since δx_j under the integral is arbitrary, the large parenthesis must vanish, and

$$\frac{\partial m}{\partial x_j} - \frac{d}{dz} \left(\frac{\partial m}{\partial x'_j} \right) = 0 \tag{2.104}$$

for $j = 1, 2$. This represents a coupled pair of Euler–Lagrange equations. They are the exact ray equations for a single particle in the case of a general curvilinear axis. They can be solved in principle for the transverse position $\mathbf{x}$ and the transverse component of the ray slope $\mathbf{x}'$ in terms of the axial coordinate z. This is a necessary condition for a path of physically allowable motion; i.e., for the path to be a ray.

The choice of coordinates x_j and slopes x'_j remains arbitrary. For example, one could choose Cartesian coordinates $x(z)$ and

$y(z)$ in the local transverse plane. Alternatively, one could treat the local transverse plane as the complex plane with coordinates $u(z) = x(z) + iy(z)$ and $\bar{u}(z) = x(z) - iy(z)$. Alternatively, one could choose polar coordinates $r(z)$ and $\theta(z)$ in the local transverse plane. The best choice is the one which allows one to express the problem in the simplest possible way.

We can write

$$
\begin{aligned}
n &= \mathbf{P} \cdot \hat{\mathbf{s}} = \mathbf{P} \cdot \frac{d\mathbf{x}}{ds} \\
&= P_x \frac{dx}{ds} + P_y \frac{dy}{ds} + P_z \frac{dz}{ds} \\
&= (P_x x' + P_y y' + P_z) \frac{dz}{ds} \\
m &= n \frac{ds}{dz} = P_x x' + P_y y' + P_z,
\end{aligned}
\tag{2.105}
$$

from which it follows

$$
\frac{\partial m}{\partial x'_j} = P_j
\tag{2.106}
$$

for $j = 1, 2$, where P_x and P_y are the transverse components of canonical momentum. The Euler-Lagrange equations can therefore be written as

$$
\frac{\partial m}{\partial x_j} - \frac{dP_j}{dt} = 0
\tag{2.107}
$$

in analogy with (2.58). Considering two rays which are infinitesimally displaced from one another, the differential in optical path between the rays is (2.103, 2.106)

$$
\delta W_{ab} = \sum_{j=1}^{2} (P_{bj}\,\delta x_{bj} - P_{aj}\,\delta x_{aj}).
\tag{2.108}
$$

In general δW_{ab} is non-zero, since the endpoints x_{aj} and x_{bj} can be independently displaced between the two rays by δx_{aj} and δx_{bj}, respectively.

Since δW_{ab} is an exact differential, it follows that

$$
P_{bj} = \frac{\partial W_{ab}}{\partial x_{bj}}, \qquad P_{aj} = -\frac{\partial W_{ab}}{\partial x_{aj}}.
\tag{2.109}
$$

Physically, this means that the transverse component of canonical momentum is perpendicular to the contour lines of constant optical path W_{ab} in any transverse plane. Again, the coordinates $x_j(z)$ and slopes $x'_j(z)$ in (2.104) should be regarded as completely general. In the following sections it will prove expedient to move freely between alternative coordinate systems.

2.4.2 Natural units

The discussion in the following few sections will be somewhat simplified by expressing the variables in alternative units, which are derived from SI units. The scalar kinetic momentum p can be written as

$$\tilde{p} = \frac{p}{mc} \geq 0. \qquad (2.110)$$

Since the total energy needs only to be expressed to within an arbitrary, additive constant, we are free to define the zero of potential energy. Let

$$T + q\phi = 0, \qquad (2.111)$$

where T is the kinetic energy, and where $q = -e$ for the electron charge. This is consistent with energy conservation in the case where the electromagnetic potentials have no explicit time dependence. Physically, the zero of potential energy is here defined at a position where the particle has zero kinetic energy, i.e., is at rest. This position might coincide with the emission surface, but this need not necessarily be the case. The quantity ϕ thus represents both the electrostatic potential, and the kinetic energy of the particle, only for this particular choice of the zero of potential energy. Many workers call ϕ the *beam voltage* at any given position in the optical system. We define a dimensionless quantity

$$\tilde{\phi} = -\frac{q\phi}{mc^2} = \frac{T}{mc^2} \geq 0. \qquad (2.112)$$

The magnetic vector potential can be written in dimensionless form as

$$\tilde{\mathbf{A}} = -\frac{q\,\mathbf{A}}{mc}.$$ (2.113)

The velocity, space charge density, and space current density can be written as

$$\tilde{\mathbf{v}} = \frac{\mathbf{v}}{c}, \qquad \tilde{\rho} = -\frac{q}{mc^2}\frac{\rho}{\epsilon_0}, \qquad \tilde{\mathbf{j}} = -\frac{q}{mc}\mu_0\,\mathbf{j}, \qquad \tilde{\mathbf{j}} = \tilde{\rho}\,\tilde{\mathbf{v}},$$ (2.114)

respectively, where $\tilde{\rho} \le 0$, regardless of sign of charge q.

The rest energy plus the kinetic energy is given in dimensionless units by

$$\gamma = 1 + \phi = \sqrt{1 + p^2},$$ (2.115)

where the rest energy mc^2 is unity in these units. Solving this for the scalar kinetic momentum p, we obtain (2.115)

$$p = \sqrt{2\phi + \phi^2},$$ (2.116)

where ϕ and p can be regarded as functions of the coordinates x_j only. This is due to the fact that the zero of potential energy is fixed (2.111). In the nonrelativistic limit, the kinetic energy is small relative to the rest mass, as follows:

$$\phi \ll 1.$$ (2.117)

In the following discussion, we will not make this approximation, but rather retain the full relativistic form throughout.

All quantities are dimensionless except coordinates and time, which retain their SI units of meters and seconds, respectively. One can easily return to SI units at any point in a calculation by inverting the above transformations. Many calculations seek position, such as the path of a ray, or the deviation of the path from its paraxial or Gaussian approximation. In such cases, it is not necessary to convert back to SI units for the result to be practical. We therefore refer to these units as *natural* units. Unless specifically noted, we will use these units throughout the following section describing the special case with axial symmetry, and drop the tilde.

2.5 Axial symmetry

Systems with a straight optic axis, where the potentials $\mathbf{A}$ and ϕ are axially symmetric, represent an important special case of the general curvilinear axis. This includes a large class of useful instruments, including electron and ion microscopes. It excludes curved-axis energy analyzers. The reader is referred to Rose [75], Hawkes and Kasper [43, 44], and Wollnik [93] for further detail and elaboration, both of axially symmetric and nonsymmetric systems.

2.5.1 Exact equations of motion for axially symmetric fields

In the absence of space charge, the electrostatic potential ϕ satisfies Laplace's equation,

$$\nabla^2 \phi = 0. \tag{2.118}$$

In cylindrical coordinates this becomes

$$\left(\frac{\partial^2}{\partial r^2} + \frac{1}{r} \frac{\partial}{\partial r} + \frac{\partial^2}{\partial z^2} \right) \phi = 0. \tag{2.119}$$

We propose a series solution by the method of undetermined coefficients [74]. We assume that ϕ can be expanded in a series representation given by

$$\phi(r, z) = a_0(z) + a_2(z)\, r^2 + a_4(z)\, r^4 + \ldots, \tag{2.120}$$

where we will now proceed to solve for the coefficients a_j. The kinetic energy of a particle on axis is

$$a_0(z) = \phi(0, z) \equiv \Phi(z). \tag{2.121}$$

From (2.119, 2.120, 2.121) it follows that

$$\phi(r, z) = \Phi - \tfrac{1}{4}\Phi'' \, r^2 + \tfrac{1}{64}\Phi^{IV} \, r^4 + \ldots, \tag{2.122}$$

where primes indicate differentiation with respect to z. Expanding the scalar kinetic momentum p we obtain (2.116, 2.122):

$$
\begin{aligned}
p(r, z) &= \sqrt{2\phi + \phi^2} \\
&= \mathsf{p} - \tfrac{1}{4}\mathsf{p}^{-1}\,\Phi''\,(1 + \Phi)\,r^2 \\
&\quad + \left[\tfrac{1}{64}\mathsf{p}^{-1}\,\Phi^{IV}\,(1 + \Phi) - \tfrac{1}{32}\mathsf{p}^{-3}\,\Phi''^{\,2}\right]\,r^4 + \ldots,
\end{aligned}
\tag{2.123}
$$

where we have defined a quantity $\mathsf{p}(z)$ as the scalar kinetic momentum on axis as follows (2.121, 2.122):

$$\mathsf{p}(z) \equiv p(0, z) = \sqrt{2\Phi + \Phi^2}. \tag{2.124}$$

Separately, the magnetic field $\mathbf{B}$ is given in terms of the magnetic vector potential $\mathbf{A}$ as

$$\mathbf{B} = \nabla \times \mathbf{A} = -\hat{\mathbf{r}}\,\frac{\partial A_\theta}{\partial z} + \hat{\mathbf{z}}\left(\frac{\partial A_\theta}{\partial r} + \frac{A_\theta}{r}\right). \tag{2.125}$$

In the absence of space current, Maxwell's equation is

$$\nabla \times \mathbf{B} = 0, \tag{2.126}$$

from which it follows that (2.125, 2.126)

$$-\hat{\theta}\left(\frac{\partial^2}{\partial r^2} + \frac{1}{r}\frac{\partial}{\partial r} - \frac{1}{r^2} + \frac{\partial^2}{\partial z^2}\right)A_\theta = 0. \tag{2.127}$$

We assume a series representation for A_θ as follows:

$$A_\theta(r, z) = b_1(z)\,r + b_3(z)\,r^3 + b_5(z)\,r^5 + \ldots, \tag{2.128}$$

where we now proceed to solve for the coefficients b_j. We define a function $\mathsf{B}(z)$ as the magnetic field on axis,

$$\mathsf{B}(z) \equiv B_z(0, z) = 2b_1 \tag{2.129}$$

leading to

$$A_\theta(r, z) = \tfrac{1}{2}\,\mathsf{B}r - \tfrac{1}{16}\,\mathsf{B}''\,r^3 + \tfrac{1}{384}\,\mathsf{B}^{IV}\,r^5 + \dots \tag{2.130}$$

The modified refractive index m is written as

$$
\begin{aligned}
m &= n\,\frac{ds}{dz} \\
&= (p - \mathbf{A} \cdot \hat{\mathbf{s}})\,\frac{ds}{dz} \\
&= p\,\sqrt{1 + r'^2 + r^2\theta'^2} - \frac{v_\theta A_\theta}{ds/dt}\,\frac{ds}{dz} \\
&= p\,\sqrt{1 + r'^2 + r^2\theta'^2} - r\theta'\,A_\theta. \tag{2.131}
\end{aligned}
$$

Euler-Lagrange equation for angular coordinate is (2.104, 2.131)

$$\frac{\partial m}{\partial \theta} - \frac{d}{dz}\frac{\partial m}{\partial \theta'} = 0, \tag{2.132}$$

where, because of axial symmetry,

$$\frac{\partial m}{\partial \theta} = 0. \tag{2.133}$$

From (2.131) we obtain

$$\frac{\partial m}{\partial \theta'} = \frac{pr^2\theta'}{\sqrt{1 + r'^2 + r^2\theta'^2}} - rA_\theta = C = const, \tag{2.134}$$

where C is identified (2.106, 2.134) as the conserved canonical angular momentum. In the case where $C = 0$, the ray intersects the optic axis at some point. Such a ray has no angular momentum, and is called a *meridional* ray. In the case where $C \neq 0$, the ray has angular momentum, and does not intersect the optic axis. Such a ray is called a *skew* ray, with C as a measure of skewness.

The Euler-Lagrange equation for the radial coordinate is (2.104)

$$\frac{\partial m}{\partial r} - \frac{d}{dz}\frac{\partial m}{\partial r'} = 0. \tag{2.135}$$

This leads (2.131, 2.135) to the exact ray equation for the radial coordinate in the case of axial symmetry as follows:

$$\frac{d}{dz}\left[r'\sqrt{\frac{p^2 - (C/r + A_\theta)^2}{1 + r'^2}}\right] = \left[\frac{1 + r'^2}{p^2 - (C/r + A_\theta)^2}\right]^{1/2}$$

$$\cdot\left[\frac{(C/r + A_\theta)^2}{r} + p\frac{\partial p}{\partial r} - \left(\frac{C}{r} + A_\theta\right)\left(\frac{A_\theta}{r} + \frac{\partial A_\theta}{\partial r}\right)\right]. \tag{2.136}$$

recalling that p is the scalar kinetic momentum (2.123), and A_θ is the magnitude of the magnetic vector potential (2.130). Both p and A_θ are assumed to be known functions of the coordinates.

The differential equations (2.134, 2.136) are a coupled pair, for which the desired solutions $r(z)$ and $\theta(z)$ are exact in principle. These equations were derived by Sturrock [86]. Because the equations are nonlinear, the solutions $r(z)$ and $\theta(z)$ cannot be expressed in a simple, closed form. Consequently, an analytical solution must rely on finding a suitable approximation. Alternatively, these equations are amenable to exact numerical solution.

2.5.2　Paraxial approximation, Gaussian optics

Assuming

$$r'^2 \ll 1, \quad C = 0 \tag{2.137}$$

in (2.136), and retaining only terms through order r, one obtains (2.123, 2.130, 2.136, 2.137) the approximation

$$\frac{2+\Phi}{1+\Phi}\,\Phi\,r'' + \Phi'r' + \tfrac{1}{2}\,\Phi''r + \frac{B^2}{4\,(1+\Phi)}\,r = 0. \qquad (2.138)$$

This is a linear second order equation for the radial position $r(z)$ of a meridional ray. It is only accurate for rays close to the optic axis, and this approximation is therefore known as the paraxial approximation. A purely electrostatic field has $B = 0$, and a purely magnetic field has $\Phi = const$.

This equation can be integrated in principle by first seeking an integrating factor. To this end we define a reduced ray [33, 71]

$$R(z) = [\,\Phi(z)\,]^{1/4}\, r(z). \qquad (2.139)$$

Substituting this into (2.138), we obtain a reduced equation

$$R''(z) + Q(z)\, R(z) = 0, \qquad (2.140)$$

where we have defined a function $Q(z)$ as

$$Q(z) = \frac{3}{16}\left(\frac{\Phi'}{\Phi}\right)^2 \frac{1+\Phi}{1+\Phi/2} + \frac{B^2}{8\,\Phi\,(1+\Phi/2)}. \qquad (2.141)$$

The region where Q is non-zero constitutes a lens, completely analogous to a lens in light optics, with the difference that the boundaries for the focusing region are not sharply delineated.

Since $Q(z)$ is positive-definite, it follows that $R''(z) \leq 0$. The reduced ray $R(z)$ therefore always bends toward the optic axis. This is not necessarily true for the actual ray $r(z)$, which can curve away from the axis within a region with electric field.

We can define a forward focal length f_+ for the lens, where rays enter parallel to the optic axis at radius $r_{-\infty}$, and exit with slope r'_∞

$$\frac{1}{f_+} = -\frac{r'_\infty}{r_{-\infty}}, \qquad \text{where} \qquad r'_{-\infty} = 0. \qquad (2.142)$$

For many systems it is the case that the radial position of the ray is roughly constant within the lens field, i.e., $R(z) \approx const.$ Such a lens is called a *thin* lens. It is also often the case that the electrostatic component of the focusing is weak or nonexistent; i.e., $r' \approx \Phi^{-1/4} R'$. Such a lens is called a *weak* lens. Using these approximations, we obtain (2.139)

$$\frac{1}{f_+} \approx -\left(\frac{\Phi_{-\infty}}{\Phi_\infty}\right)^{1/4} \frac{R'_\infty}{R_{-\infty}}. \tag{2.143}$$

From (2.140) we obtain

$$R'_\infty = \int_{-\infty}^\infty R'' \, dz = -\int_{-\infty}^\infty Q \, R \, dz \approx -R_{-\infty} \int_{-\infty}^\infty Q \, dz. \tag{2.144}$$

From (2.141, 2.144, 2.145) we obtain

$$\frac{1}{f_+} \approx \left(\frac{\Phi_{-\infty}}{\Phi_\infty}\right)^{1/4} \int_{-\infty}^\infty \left[\frac{3}{16}\left(\frac{\Phi'}{\Phi}\right)^2 \frac{1+\Phi}{1+\Phi/2} + \frac{\mathsf{B}^2}{8\,\Phi\,(1+\Phi/2)}\right] dz, \tag{2.145}$$

where the first term on the right represents the electrostatic focusing, and the second term represents the magnetic focusing. Similarly, we define a reverse focal length f_-, where rays enter parallel to the optic axis at radius r_∞, and exit with slope $r'_{-\infty}$:

$$\frac{1}{f_-} = \frac{r'_{-\infty}}{r_\infty}, \qquad \text{where} \qquad r'_\infty = 0. \tag{2.146}$$

The axial positions of principal planes follow directly from f_+ and f_-.

The quantity $1/f$ represents the focal strength of a lens. In the purely electrostatic case where $\mathsf{B} = 0$, the focal strength is proportional to the charge q, and independent of the mass m, taking account of the dimensionless units. In the purely magnetic case where $\Phi' = 0$, the focal strength is proportional to the ratio of q/m. Consequently, it is more efficient to use electrostatic lenses for heavier particles, such as ions, and magnetic lenses for lighter particles, such as electrons.

2.5.3 Series solution for the general ray equation

We now seek a general solution to the exact ray equation (2.104). This must include all rays, including meridional and skew rays. Because the exact trajectory equations (2.134, 2.136) are nonlinear, they cannot be solved in closed form. Consequently, we seek an approximate solution by series expansion [33].

Recalling the modified refractive index for the general curvilinear axis,

$$m = p \sqrt{1 + r'^2 + r^2\theta'^2} - r\theta' A_\theta, \qquad (2.147)$$

where the scalar kinetic momentum $p(r, z)$ is given by (2.123) and the magnetic vector potential A_θ is given by (2.130).

We define a complex transverse coordinate

$$u = X + iY = r\,e^{i\theta}, \qquad \bar{u} = X - iY = r\,e^{-i\theta}, \qquad (2.148)$$

where $X(z)$ and $Y(z)$ are Cartesian coordinates in a transverse plane at axial coordinate z. It follows that

$$i\,(\bar{u}'\,u - \bar{u}u') = 2(X\,Y' - X'\,Y) = 2\,r^2\theta', \qquad (2.149)$$

and

$$\sqrt{1 + r'^2 + r^2\theta'^2} = \sqrt{1 + \bar{u}'\,u'} = 1 + \tfrac{1}{2}\,\bar{u}'u' - \tfrac{1}{8}\,\bar{u}'^2 u'^2 + \dots. \qquad (2.150)$$

We can write a power series expansion for the refractive index, making use of the axial symmetry of the scalar kinetic momentum $p(r, z)$ and the magnetic vector potential $A_\theta(r, z)$ in (2.123, 2.130) as follows:

$$
\begin{aligned}
m \;=\;& m_0 + m_2 + m_4 + \dots \\
=\;& \mathsf{p} \\
& + \left[-\tfrac{1}{4}\mathsf{p}^{-1}\Phi''(1 + \Phi) \right] \bar{u}u \\
& + \left[\tfrac{1}{2}\mathsf{p} \right] \bar{u}'u'
\end{aligned}
$$

$$+\left[-\tfrac{1}{4}\,\mathsf{B}\right] i\left(\bar{u}'u - \bar{u}u'\right)$$

$$+\left[-\tfrac{1}{8}\,\mathsf{p}\right] \bar{u}'^{2}u'^{2}$$

$$+\left[-\tfrac{1}{8}\,\mathsf{p}^{-1}\,\Phi''\left(1+\Phi\right)\right] \bar{u}u\bar{u}'u'$$

$$+\left[\tfrac{1}{64}\,\mathsf{p}^{-1}\,\Phi^{IV}\left(1+\Phi\right) - \tfrac{1}{32}\,\mathsf{p}^{-3}\,\Phi''^{2}\right] \bar{u}^{2}u^{2}$$

$$+\left[\tfrac{1}{32}\,\mathsf{B}''\right] i\left(\bar{u}'u - \bar{u}u'\right)\bar{u}u$$

$$+\dots, \tag{2.151}$$

where the various orders of m are defined by the power of the co-ordinates and slope components. The quantities in square brackets depend only on the fields on axis, embodied in $\Phi(z)$ and $\mathsf{B}(z)$.

The paraxial term is given by (2.151)

$$m_2 = -\tfrac{1}{4}\,\mathsf{p}^{-1}\,\Phi''\left(1+\Phi\right)\bar{u}u + \tfrac{1}{2}\,\mathsf{p}\,\bar{u}'u' - \tfrac{1}{4}\,\mathsf{B}\,i\left(\bar{u}'u - \bar{u}u'\right). \tag{2.152}$$

We now define the paraxial approximation by retaining only terms through order m_2 in (2.104, 2.151). The paraxial ray equation is then given by

$$\frac{\partial m_2}{\partial \bar{u}} - \frac{d}{dz}\frac{\partial m_2}{\partial \bar{u}'} = 0. \tag{2.153}$$

Substituting (2.152) into (2.153) we obtain

$$\frac{d}{dz}\left(\mathsf{p}\,u'\right) - i\,\mathsf{B}\,u' + \tfrac{1}{2}\left[\mathsf{p}^{-1}\,\Phi''\left(1+\Phi\right) - i\,\mathsf{B}'\right]u = 0. \tag{2.154}$$

The imaginary terms correspond physically to a rotation of the bundle of rays about the optic axis as a function of the axial co-ordinate z. Physically, this arises from the Lorentz force (2.15), where the axial component of the magnetic field acts on the transverse component of the particle velocity.

It is possible to rotate the coordinate system to compensate for this. We define a rotated complex coordinate $v(z) = x(z) + i\,y(z)$ as follows:

$$u(z) = v(z)\,e^{i\chi(z)}, \tag{2.155}$$

where, by definition,

$$\frac{d\chi}{dz} = \tfrac{1}{2}\,\mathsf{p}^{-1}\,\mathsf{B}. \tag{2.156}$$

The rotation angle is then given by

$$\chi_{ab} = \tfrac{1}{2} \int_{z_a}^{z_b} \mathsf{p}^{-1} \, \mathsf{B} \, dz. \tag{2.157}$$

This gives rise to the following useful transformations:

$$
\begin{aligned}
\bar{u}\,u &= \bar{v}\,v \\
\bar{u}'u' &= \bar{v}'v' + \tfrac{1}{2}\mathsf{p}^{-1}\,\mathsf{B}\,i\,(\bar{v}'v - \bar{v}\,v') + \tfrac{1}{4}\mathsf{p}^{-2}\,\mathsf{B}^2\,\bar{v}\,v \\
i\,(\bar{u}'u - \bar{u}\,u') &= i\,(\bar{v}'v - \bar{v}\,v') + \mathsf{p}^{-1}\,\mathsf{B}\,\bar{v}\,v.
\end{aligned} \tag{2.158}
$$

Substituting these into (2.151) we obtain the modified refractive index in the rotated coordinates as

$$
\begin{aligned}
m &= m_0 + m_2 + m_4 + \ldots \\
&= \mathsf{p} \\
&\quad + \left[\tfrac{1}{2}\mathsf{p}\right] \bar{v}'v' \\
&\quad + \left[-\tfrac{1}{4}\mathsf{p}^{-1}\Phi''(1+\Phi) - \tfrac{1}{8}\mathsf{p}^{-1}\mathsf{B}^2\right]\bar{v}\,v \\
&\quad + \left[\tfrac{1}{64}\mathsf{p}^{-1}\Phi^{IV}(1+\Phi) - \tfrac{1}{32}\mathsf{p}^{-3}\Phi''^2 + \tfrac{1}{32}\mathsf{p}^{-1}\mathsf{B}\,\mathsf{B}''\right. \\
&\qquad \left. - \tfrac{1}{128}\mathsf{p}^{-3}\mathsf{B}^4 - \tfrac{1}{32}\mathsf{p}^{-3}\Phi''(1+\Phi)\mathsf{B}^2\right]\bar{v}^2\,v^2 \\
&\quad + \left[-\tfrac{1}{8}\mathsf{p}^{-1}\,\Phi''(1+\Phi) - \tfrac{1}{16}\mathsf{p}^{-1}\mathsf{B}^2\right]\bar{v}\,v\,\bar{v}'\,v' \\
&\quad + \left[-\tfrac{1}{8}\mathsf{p}\right]\bar{v}'^2\,v'^2 \\
&\quad + \left[-\tfrac{1}{32}\mathsf{p}^{-2}\mathsf{B}^3 - \tfrac{1}{16}\mathsf{p}^{-2}\Phi''(1+\Phi)\mathsf{B} + \tfrac{1}{32}\mathsf{B}''\right] \\
&\qquad \cdot\, i\,(\bar{v}'v - \bar{v}\,v')\,\bar{v}\,v \\
&\quad + \left[-\tfrac{1}{8}\mathsf{B}\right]i\,(\bar{v}'v - \bar{v}\,v')\,\bar{v}'\,v' \\
&\quad + \left[-\tfrac{1}{32}\mathsf{p}^{-1}\mathsf{B}^2\right]\left[i\,(\bar{v}'v - \bar{v}\,v')\right]^2 \\
&\quad + \ldots.
\end{aligned} \tag{2.159}
$$

The paraxial term is given in the rotated system by

$$m_2 = \tfrac{1}{2}\mathsf{p}\,\bar{v}'v' + \left[-\tfrac{1}{4}\mathsf{p}^{-1}\,\Phi''(1+\Phi) - \tfrac{1}{8}\mathsf{p}^{-1}\mathsf{B}^2\right]\bar{v}v. \tag{2.160}$$

The paraxial approximation to (2.104) is then

$$\frac{\partial m_2}{\partial \bar{v}} - \frac{d}{dz}\frac{\partial m_2}{\partial \bar{v}'} = 0. \tag{2.161}$$

Substituting (2.160) into (2.161) we obtain the paraxial ray equation in the rotated system as follows:

$$\frac{d}{dz}\left(\mathsf{p}\,v'\right) + \left[\tfrac{1}{2}\mathsf{p}^{-1}\,\Phi''\left(1+\Phi\right) + \tfrac{1}{4}\mathsf{p}^{-1}\,\mathsf{B}^2\right] v = 0. \qquad (2.162)$$

The absence of imaginary terms shows that the rotation has been removed. It is simpler to work in the rotated system than the unrotated system, as the image has the same rotation as the object in the rotated coordinates.

As a second-order linear differential equation, (2.162) has two linearly independent solutions, which we denote $g(z)$ and $h(z)$. By substituting these in turn into (2.162) for $v(z)$ and subtracting the two equations, it is straightforward to show that

$$h\,\frac{d}{dz}\left(\mathsf{p}\,g'\right) - g\,\frac{d}{dz}\left(\mathsf{p}\,h'\right) = 0, \qquad (2.163)$$

from which it follows that

$$\frac{d}{dz}\left[\mathsf{p}\left(g\,h' - g'\,h\right)\right] = 0. \qquad (2.164)$$

The quantity in square brackets is, therefore, conserved. We denote this quantity as k, defined as

$$k = \mathsf{p}(z)\left[\,g(z)\,h'(z) - g'(z)\,h(z)\,\right] = const. \qquad (2.165)$$

The conserved quantity k is called the *Wronskian*. A more general expression for the Wronskian exists for a general curvilinear axis. The reader is referred to Rose [75] for details. It is closely related to the Lagrange invariant discussed earlier.

In order to fully determine the solutions $g(z)$ and $h(z)$, it is necessary to specify boundary conditions. We choose these arbitrarily as

$$\begin{aligned} g(z_O) &= 1, & g(z_A) &= 0, \\ h(z_O) &= 0, & h(z_A) &= 1, \end{aligned} \qquad (2.166)$$

where z_O and z_A are the axial coordinates of the object and aperture planes, respectively. Given this, a general solution for the paraxial ray $v(z)$ can be written as

$$\begin{aligned} v(z) &= v_O \, g(z) + v_A \, h(z) \\ v'(z) &= v_O \, g'(z) + v_A \, h'(z). \end{aligned} \qquad (2.167)$$

Remembering $v = x + iy$ in the rotated system, it follows that

$$\begin{aligned} x_j(z) &= x_{Oj} \, g(z) + x_{Aj} \, h(z) \\ x'_j(z) &= x_{Oj} \, g'(z) + x_{Aj} \, h'(z), \end{aligned} \qquad (2.168)$$

where $x_j = (x, y)$ for $j = (1, 2)$.

The choice of aperture plane z_A is arbitrary. Often one chooses the aperture plane to coincide with a physical aperture, but this need not be the case. The aperture plane cannot coincide with the object or image planes, as the solutions g and h would no longer be independent.

Since the Wronskian is conserved, it retains the same value throughout the system, and

$$k = \mathsf{p}_O \, h'_O = -\mathsf{p}_A \, g'_A = \mathsf{p}_I \, M \, h'_I, \qquad (2.169)$$

where the subscripts O, A, and I denote the object plane, aperture plane, and Gaussian image plane, respectively, and M is the magnification.

In practice one usually determines the solutions $g(z)$ and $h(z)$ by solving (2.162) numerically. This requires knowing $\Phi''(z)$ to high accuracy. Unfortunately this is not always possible, even if one knows $\Phi(z)$ to high accuracy at discrete points along the axis, because numerical differentiation introduces error. The dependence on the second-order derivative can be eliminated by defining a reduced ray $w(z)$ as follows:

$$v(z) = [\,\mathsf{p}(z)\,]^{-1/2} \, w(z). \qquad (2.170)$$

From (2.160, 2.170) we obtain

$$
\begin{aligned}
m_2 = {} & \tfrac{1}{2}\bar{w}'w' \\
& + \left[-\tfrac{1}{4}\mathsf{p}^{-2}\,\Phi''\,(1+\Phi) + \tfrac{1}{8}\mathsf{p}^{-4}\,\Phi'^{2}\,(1+\Phi)^{2} - \tfrac{1}{8}\mathsf{p}^{-2}\mathsf{B}^{2} \right]\bar{w}w \\
& - \tfrac{1}{4}\mathsf{p}^{-2}\,\Phi'\,(1+\Phi)\,(\bar{w}'w + \bar{w}w'). \quad\quad (2.171)
\end{aligned}
$$

From (2.104) we have

$$
\frac{\partial m_2}{\partial \bar{w}} - \frac{d}{dz}\frac{\partial m_2}{\partial \bar{w}'} = 0. \quad\quad (2.172)
$$

This yields the paraxial ray equation in reduced coordinates as follows:

$$
\frac{d^{2}w}{dz^{2}} + \left[\tfrac{3}{4}\mathsf{p}^{-4}\,\Phi'^{2}\,(1+\Phi)^{2} - \tfrac{1}{2}\mathsf{p}^{-2}\,\Phi'^{2} + \tfrac{1}{4}\mathsf{p}^{-2}\mathsf{B}^{2} \right]w = 0, \quad (2.173)
$$

where the second-order derivative Φ'' has been successfully eliminated. Although the reduced ray $w(z)$ offers a practical simplification for obtaining the paraxial ray solution numerically, it offers no advantage for obtaining the aberrations.

Axial symmetry permits only terms (2.151, 2.159) in m of the form

$$
\begin{array}{cccc}
X^{2}+Y^{2} & X'^{2}+Y'^{2} & 2\,(XY'-X'Y) & 2\,(XX'+YY') \\
\bar{u}\,u & \bar{u}'\,u' & i(\bar{u}'u - \bar{u}u') & \bar{u}'u + \bar{u}u' \\
x^{2}+y^{2} & x'^{2}+y'^{2} & 2\,(xy'-x'y) & 2\,(xx'+yy') \\
\bar{v}\,v & \bar{v}'\,v' & i(\bar{v}'v - \bar{v}v') & \bar{v}'v + \bar{v}v' \\
\bar{w}\,w & \bar{w}'\,w' & i(\bar{w}'w - \bar{w}w') & \bar{w}'w + \bar{w}w'.
\end{array}
$$

This is shown by replacing $x \to y$ and $y \to -x$, for example, corresponding to a rotation of the coordinate system by $+90$ degrees in the transverse plane. The reader can easily verify that all of the above products are invariant under all such 90 degree rotations. Exactly four independent degrees of freedom exist, corresponding to two transverse coordinates and two transverse slope components. As a result, four independent products exist for each line of the above table. These facts represent the necessary and sufficient conditions for axial symmetry.

We have derived a prescription for obtaining the general solution in the paraxial approximation. The linearity of the paraxial ray equation ensures perfect imaging in this approximation. Departure from perfect imaging represents aberration. This will be treated in a later section using a perturbation approach. First, however, it is instructive to extend the preceding arguments to include the effects of space charge. This is the subject of the next section.

2.5.4   Space charge

In the classical limit, particles in the beam can be regarded as discrete, point charges. The particles propagate together, each with its own velocity. If the beam is sufficiently dense, the particles interact with one another via the Lorentz force (2.15). Every particle produces an electrostatic field **E** by virtue of its charge, and a magnetic field **B** by virtue of its current. These fields, in turn, act on the other particles in the beam.

A proper analysis in the classical limit treats the particles as discrete, and randomly distributed within the beam. This will be done in the later section on the stochastic interaction. A great deal of understanding can be gained by regarding the beam as a continuum of charge and current, however. We consider the effect of the fields generated on a test particle which moves with the beam. We imagine an axially symmetric, monoenergetic beam characterized by space charge density $\rho(r)$ and current density $j(r)$. The geometry in the lab frame is shown schematically in Figure 2.7. Initially the beam is assumed to be parallel, in that the direction of the local current density vector throughout the beam cross section points everywhere along the z-axis at the left of the figure.

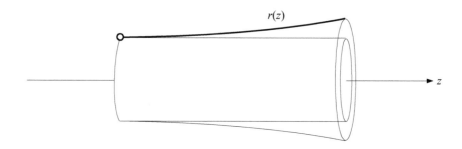

Figure 2.7: Space charge, quasi-parallel beam.

Considering a cylinder of radius r, the electric field vector $\mathbf{E}$ points radially outward for a positively charged beam, and radially inward for a negatively charged beam. Gauss's law can be written in SI units as

$$\int_S \mathbf{E} \cdot d\mathbf{S} = \frac{1}{\epsilon_0} \int_V \rho(r)\, dV, \tag{2.174}$$

where the volume integral on the right is the total charge enclosed witbn the cylinder. Expressing the elements of surface area and volume in cylindrical coordinates, this becomes

$$E_r(r) = \frac{1}{\epsilon_0 r} \int_0^r \rho(r_1)\, r_1 \, dr_1. \tag{2.175}$$

The ends of the cylinder do not contribute, because the electric field vector is coplanar with the end faces. Any axially symmetric charge outside of the radius r does not contribute to the electric field. Separately, Ampere's law can be written in SI units as

$$\int \mathbf{B} \cdot d\mathbf{l} = \mu_0 \int j \, dA, \tag{2.176}$$

where the integral on the left is the line integral around a circular path of radius r, and the integral on the right is the area integral over a circular disk, which is oriented in the transverse plane. The integral on the right side is the total current enclosed within the cylinder. Expressing the elements of path length and transverse area in polar coordinates, this becomes

$$B_\theta(r) = \frac{\mu_0}{r} \int_0^r j(r_1)\, r_1 \, dr_1. \tag{2.177}$$

We now consider the Lorentz force (2.15) acting on a test particle of charge q at radius r which moves with the beam. This particle is depicted schematically by the small circle in Figure 2.7. The radial component of the Lorentz force can be written as

$$\gamma \, m \frac{d^2 r}{dt^2} = \frac{q}{\epsilon_0 v \, r} \left(1 - \frac{v^2}{c^2}\right) \int_0^r j(r_1)\, r_1 \, dr_1, \tag{2.178}$$

where γ is defined by (2.9), and we have made use of

$$j(r) = \rho(r)\, v, \tag{2.179}$$

and

$$\mu_0 \, \epsilon_0 = 1/c^2. \tag{2.180}$$

The first term in large parentheses represents the outwardly directed electrostatic force arising from the space charge ρ, while the second term represents the inwardly directed magnetic force arising from the space current j. The relative strength of these two forces approaches equality in the extreme relativistic limit where $v^2/c^2 \to 1$. Physically, this occurs because the interaction time approaches zero in the lab frame. The presence of opposing electrostatic and magnetic forces can therefore be regarded as a purely relativistic effect. We now write

$$\frac{dr}{dt} = \frac{dr}{dz}\frac{dz}{dt} = v\, r', \qquad \frac{d^2 r}{dt^2} = v^2\, r'', \tag{2.181}$$

where primes denote differentiation with respect to the axial coordinate z. This leads immediately to

$$r''(z) = \frac{q\, m^2}{\epsilon_0\, p^3\, r} \int_0^r j(r_1)\, r_1 \, dr_1, \tag{2.182}$$

where p is the relativistic scalar kinetic momentum obeying (2.28, 2.20).

$$p = \beta\, \gamma\, mc. \tag{2.183}$$

The momentum p is constant to first order in this approximation. The momentum p is related to the kinetic energy T and relativistic beam voltage V^* by (2.96). The equation (2.182) can be regarded as a general differential-integral equation for the trajectory $r(z)$ for a quasi-parallel beam in the first-order (paraxial) continuum approximation. It is possible in principle to solve this equation for the trajectory $r(z)$ of the test particle. This trajectory is shown schematically as the bold curve in Figure 2.7. This curve traces out the envelope of the expanding beam.

In order to gain a further appreciation of the physical significance of this, we consider the special case where the current density j is constant within the volume of the beam. We can write $j(r) = j_0$, independent of radius r. We further assume that the effect is weak, such that the expansion of the beam is small relative to the beam radius. In this case the equation (2.182) reduces to

$$r''(z) = \left(\frac{q\, m^2\, j_0}{2\, \epsilon_0 p^3} \right) r(z). \tag{2.184}$$

The leading factor in large parentheses on the right side can be regarded as constant in this approximation. Physically, the left side represents the bending of the ray. This is proportional to the distance r off axis. This is precisely the condition for a perfect lens, with the result that defocusing occurs, but no blurring. This defocus can be corrected in principle, and has no net effect on the quality of the image.

With this intuitive picture in mind, we now proceed to apply the methods of the preceding sections, taking the average space charge and space current into account. We return to dimensionless units (2.110 to 2.117) at this point. The electrostatic potential in the presence of space charge obeys Poisson's equation,

$$\nabla^2 \phi = -\rho. \tag{2.185}$$

We assume the space charge $\rho = \rho(z)$ to be uniform in the transverse plane while varying in the axial direction. From (2.120, 2.121,

2.185),

$$\phi(r,z) = \Phi - \tfrac{1}{4}(\Phi'' + \rho)\,r^2 + \tfrac{1}{64}(\Phi^{IV} + \rho'')\,r^4 + \dots \qquad (2.186)$$

The scalar kinetic momentum is (2.116, 2.186)

$$\begin{aligned}
p &= \mathsf{p} - \tfrac{1}{4}\mathsf{p}^{-1}(\Phi'' + \rho)(1 + \Phi)\,r^2 \\
&\quad + \tfrac{1}{64}\mathsf{p}^{-1}(\Phi^{IV} + \rho'')(1 + \Phi)\,r^4 - \tfrac{1}{32}\mathsf{p}^{-3}(\Phi'' + \rho)^2\,r^4 + \dots
\end{aligned}$$

$$(2.187)$$

In addition to the space charge, a beam in general has a global (averaged) space current $\mathbf{j}$. Maxwell's equation is

$$\nabla \times \mathbf{B} = \nabla \times (\nabla \times \mathbf{A}) = \mathbf{j}. \qquad (2.188)$$

In the lab frame, the beam current is primarily along the beam axis, with the transverse component relatively small. We therefore neglect the transverse components of $\mathbf{j}$ and $\mathbf{A}$. It follows that (2.188):

$$j_z = -\frac{\partial^2 A_z}{\partial r^2} - \frac{1}{r}\frac{\partial A_z}{\partial r}. \qquad (2.189)$$

The solution for the axial component of the magnetic vector potential A_z arising from the space current is

$$A_z = -\tfrac{1}{4}j_z\,r^2 = -\tfrac{1}{4}\frac{\mathsf{p}}{1 + \Phi}\rho\,r^2, \qquad (2.190)$$

where the space current and space charge are related by

$$j_z = \rho\,v_z = \frac{\mathsf{p}}{1 + \Phi}\rho. \qquad (2.191)$$

Using the earlier approach, we obtain the modified refractive index (2.147) as follows:

$$m = p\sqrt{1 + r'^2 + r^2\theta'^2} - r\theta'\,A_\theta - A_z. \qquad (2.192)$$

The expression of (2.159) with space charge terms added to (2.192) gives the result

$$m = m_0 + m_2 + m_4 + \dots$$

$$
\begin{aligned}
= \; & \mathsf{p} \\
& + \left[\tfrac{1}{2}\,\mathsf{p} \right] \bar{v}'\,v' \\
& + \left[-\tfrac{1}{4}\,\mathsf{p}^{-1} \left(\Phi'' + \frac{\rho}{(1+\Phi)^2} \right) (1+\Phi) - \tfrac{1}{8}\,\mathsf{p}^{-1}\mathsf{B}^2 \right] \bar{v}\,v \\
& + \big[\, \tfrac{1}{64}\,\mathsf{p}^{-1}\left(\Phi^{IV} + \rho'' \right)(1+\Phi) - \tfrac{1}{32}\,\mathsf{p}^{-3}\left(\Phi'' + \rho \right)^2 \\
& \qquad + \tfrac{1}{32}\,\mathsf{p}^{-1}\mathsf{B}\,\mathsf{B}'' - \tfrac{1}{128}\,\mathsf{p}^{-3}\mathsf{B}^4 \\
& \qquad - \tfrac{1}{32}\,\mathsf{p}^{-3}(\Phi'' + \rho)(1+\Phi)\,\mathsf{B}^2 \big]\, \bar{v}^2\,v^2 \\
& + \left[-\tfrac{1}{8}\,\mathsf{p}^{-1}\left(\Phi'' + \rho \right)(1+\Phi) - \tfrac{1}{16}\,\mathsf{p}^{-1}\mathsf{B}^2 \right] \bar{v}\,v\,\bar{v}'\,v' \\
& + \left[-\tfrac{1}{8}\,\mathsf{p} \right] \bar{v}'^{\,2}\,v'^{\,2} \\
& + \left[-\tfrac{1}{32}\,\mathsf{p}^{-2}\mathsf{B}^3 - \tfrac{1}{16}\,\mathsf{p}^{-2}(\Phi'' + \rho)(1+\Phi)\,\mathsf{B} + \tfrac{1}{32}\,\mathsf{B}'' \right] \\
& \qquad \cdot\, i\,(\bar{v}'\,v - \bar{v}\,v')\,\bar{v}\,v \\
& + \left[-\tfrac{1}{8}\,\mathsf{B} \right] i\,(\bar{v}'\,v - \bar{v}\,v')\,\bar{v}'\,v' \\
& + \left[-\tfrac{1}{32}\,\mathsf{p}^{-1}\mathsf{B}^2 \right] \left[i\,(\bar{v}'\,v - \bar{v}\,v') \right]^2 \\
& + \dots
\end{aligned}
\tag{2.193}
$$

We note that the paraxial space charge term (large parentheses) tends to zero in the extreme relativistic limit $\Phi \gg 1$. Physically, this occurs because the space current causes magnetic compression of the beam, due to parallel current elements. This purely relativistic effect offsets the Coulomb repulsion of the space charge. Equivalently, the interaction time approaches zero in the lab frame, causing the space charge interaction to approach zero as well.

2.5.5   The primary geometrical aberrations

We showed previously that the exact ray equation for the case of axial symmetry cannot be solved analytically in closed form. Consequently, we adopted a series solution. The paraxial approximation leads to a linear, second order differential equation for the ray, which can be solved in principle, but is only accurate for rays near the optic axis. The next step is to solve for the aberrations. This requires the next approximation beyond the paraxial approximation. The following analysis closely follows that of Glaser [33], and in addition properly includes the effects of special relativity. This is also treated in detail by Rose [75].

We begin by defining the action integral (mechanical analog of the light-optical path length) W_2 between any two planes z_a and z_b in the paraxial approximation,

$$W_2 = \int_{z_a}^{z_b} m_2 \, dz, \qquad (2.194)$$

where m_2 is assumed to be known from the preceding analysis (2.159). The paraxial ray equation is

$$\frac{\partial m_2}{\partial x_j} - \frac{d}{dz} \frac{\partial m_2}{\partial x'_j} = 0. \qquad (2.195)$$

The solution for the transverse Cartesian coordinates $x_j(z)$ and canonical momentum components $P_j(z)$ in the paraxial approximation is (2.168)

$$\begin{aligned} x_j(z) &= x_{Oj} \, g(z) + x_{Aj} \, h(z) \\ P_j(z) &= \mathsf{p}(z) \left[x_{Oj} \, g'(z) + x_{Aj} \, h'(z) \right]. \end{aligned} \qquad (2.196)$$

where we have made use of (2.106, 2.159) and $v(z) = x(z) + i\,y(z)$ to obtain

$$P_j(z) = \mathsf{p}(z)\, x'_j(z) \qquad (2.197)$$

in the paraxial approximation.

To obtain the aberration, we form the first order perturbation on the paraxial ray (2.108),

$$\delta W_2 = \int_{z_a}^{z_b} (\delta m_2) \, dz = \sum_{j=1}^{2} (P_{bj}\, \delta x_{bj} - P_{aj}\, \delta x_{aj}), \qquad (2.198)$$

where the start plane z_a and the end plane z_b are assumed to be fixed. From this it follows that

$$\delta x_{bj} = \frac{\partial}{\partial P_{bj}}\,(\delta W_2), \qquad \delta x_{aj} = -\frac{\partial}{\partial P_{aj}}\,(\delta W_2), \qquad (2.199)$$

where we designate δx_{aj} and δx_{bj} as the primary aberration at the start plane z_a and end plane z_b respectively. We invert the paraxial solution (2.196) to solve for (x_{Oj}, x_{Aj}) in terms of (x_j, P_j):

$$\begin{aligned}
x_{Oj} &= k^{-1}\,(\mathsf{p}\,h'\,x_j - h\,P_j\,) \\
x_{Aj} &= k^{-1}\,(g\,P_j - \mathsf{p}\,g'\,x_j\,),
\end{aligned} \qquad (2.200)$$

where k is the conserved Wronskian (2.165). From the chain rule for partial derivatives,

$$\begin{aligned}
\frac{\partial}{\partial P_j}\,(\delta W_2) &= \left(\frac{\partial x_{Oj}}{\partial P_j}\frac{\partial}{\partial x_{Oj}} + \frac{\partial x_{Aj}}{\partial P_j}\frac{\partial}{\partial x_{Aj}}\right)(\delta W_2) \\
&= \left(-\frac{h}{k}\frac{\partial}{\partial x_{Oj}} + \frac{g}{k}\frac{\partial}{\partial x_{Aj}}\right)(\delta W_2). \qquad (2.201)
\end{aligned}$$

We now identify the start plane with the object plane, $z_a = z_O$, and we identify the end plane with the Gaussian image plane, $z_b = z_I$. From the paraxial solution we have, by definition

$$h(z_I) = 0, \qquad g(z_I) = \mathrm{M}, \qquad (2.202)$$

where M is the magnification. The primary aberration at the Gaussian image plane is then (2.199, 2.201)

$$\delta x_{Ij} = \frac{\mathrm{M}}{k}\frac{\partial}{\partial x_{Aj}}\int_{z_O}^{z_I} m_4\,dz, \qquad (2.203)$$

where we have identified the perturbation $\delta m_2 = m_4$. From (2.159) we can express the perturbation m_4 in the compact series form as

$$\begin{aligned}
m_4 = \;& L\,(x^2 + y^2)^2 + M\,(x^2 + y^2)\,(x'^2 + y'^2) + N\,(x'^2 + y'^2)^2 \\
&+ P\cdot 2\,(xy' - x'y)\,(x^2 + y^2) + Q\cdot 2\,(xy' - x'y)\,(x'^2 + y'^2) \\
&+ K\,[\,2\,(xy' - x'y)\,]^2, \qquad (2.204)
\end{aligned}$$

where we have substituted

$$
\begin{aligned}
\bar{v}\,v &= x^2 + y^2 \\
\bar{v}'\,v' &= x'^2 + y'^2 \\
i\,(\bar{v}'\,v - \bar{v}\,v') &= 2\,(x\,y' + x'\,y) \tag{2.205}
\end{aligned}
$$

in (2.159). We have defined field coefficients (2.159, 2.204) in natural units as

$$
\begin{aligned}
L &= \tfrac{1}{64}\,\mathsf{p}^{-1}\Phi^{IV}(1+\Phi) - \tfrac{1}{32}\,\mathsf{p}^{-3}\Phi''^{\,2} + \tfrac{1}{32}\,\mathsf{p}^{-1}\mathsf{B}\,\mathsf{B}'' - \tfrac{1}{128}\,\mathsf{p}^{-3}\mathsf{B}^4 \\
&\quad - \tfrac{1}{32}\,\mathsf{p}^{-3}\Phi''(1+\Phi)\,\mathsf{B}^2 \\
M &= -\tfrac{1}{8}\,\mathsf{p}^{-1}\,\Phi''(1+\Phi) - \tfrac{1}{16}\,\mathsf{p}^{-1}\mathsf{B}^2 \\
N &= -\tfrac{1}{8}\,\mathsf{p} \\
P &= -\tfrac{1}{32}\,\mathsf{p}^{-2}\mathsf{B}^3 - \tfrac{1}{16}\,\mathsf{p}^{-2}\Phi''\,(1+\Phi)\,\mathsf{B} + \tfrac{1}{32}\,\mathsf{B}'' \\
Q &= -\tfrac{1}{8}\,\mathsf{B} \\
K &= -\tfrac{1}{32}\,\mathsf{p}^{-1}\mathsf{B}^2, \tag{2.206}
\end{aligned}
$$

remembering the definition (2.124) for the scalar kinetic momentum on axis p. The effects of uniform space charge density $\rho(z)$ can be included (2.193) by substituting

$$
\begin{aligned}
\Phi'' &\to \Phi'' + \rho \\
\Phi^{IV} &\to \Phi^{IV} + \rho'' \tag{2.207}
\end{aligned}
$$

in the field coefficients L, M, and P above (2.206). The solution (2.168) for the paraxial ray $x_j(z)$ is

$$
\begin{aligned}
x(z) &= x_O\,g(z) + x_A\,h(z) \\
y(z) &= y_O\,g(z) + y_A\,h(z)
\end{aligned}
$$

$$
\begin{aligned}
x'(z) &= x_O\,g'(z) + x_A\,h'(z) \\
y'(z) &= x_O\,g'(z) + x_A\,h'(z). \tag{2.208}
\end{aligned}
$$

Following Glaser, we define a new variable set

$$
\begin{aligned}
R &= x_O{}^2 + y_O{}^2 \\
\rho &= x_A{}^2 + y_A{}^2 \\
\chi &= x_O\,x_A + y_O\,y_A \\
\sigma &= x_O\,y_A - y_O\,x_A. \tag{2.209}
\end{aligned}
$$

It follows (2.208, 2.209) that

$$\begin{aligned}
x^2 + y^2 &= R\,g^2 + \rho\,h^2 + 2\,\chi\,g\,h \\
x'^2 + y'^2 &= R\,g'^2 + \rho\,h'^2 + 2\,\chi\,g'\,h' \\
2\,(x\,y' - x'\,y) &= 2\,(g\,h' - g'\,h)\,\sigma = 2\,\mathsf{p}^{-1}\,k\,\sigma.
\end{aligned} \qquad (2.210)$$

From (2.204, 2.210) we express the perturbation m_4 in terms of the new variable set as

$$\begin{aligned}
m_4 = \quad & L\,(R\,g^2 + \rho\,h^2 + 2\,\chi\,g\,h)^2 \\
+ \quad & M\,(R\,g^2 + \rho\,h^2 + 2\,\chi\,g\,h)\,(R\,g'^2 + \rho\,h'^2 + 2\,\chi\,g'\,h') \\
+ \quad & N\,(R\,g'^2 + \rho\,h'^2 + 2\,\chi\,g'\,h')^2 \\
+ \quad & P\,(2\,\sigma\,k\,\mathsf{p}^{-1})\,(R\,g^2 + \rho\,h^2 + 2\,\chi\,g\,h) \\
+ \quad & Q\,(2\,\sigma\,k\,\mathsf{p}^{-1})\,(R\,g'^2 + \rho\,h'^2 + 2\,\chi\,g'\,h') \\
+ \quad & K\,[\,4\,\sigma^2\,k^2\,\mathsf{p}^{-2}\,(\rho\,R - \chi^2)\,],
\end{aligned} \qquad (2.211)$$

where we have made use in the K-term of the readily verifiable fact that (2.209)

$$\sigma^2 = \rho\,R - \chi^2. \qquad (2.212)$$

Expanding (2.211) and collecting terms, we now form

$$\begin{aligned}
\int_{z_O}^{z_I} m_4\,dz \;=\; & A\,R^2 + B\,\rho^2 + C\,\chi^2 + D\,R\,\rho + E\,R\,\chi \\
& +F\,\rho\,\chi + e\,R\,\sigma + f\,\rho\,\sigma + c\,\chi\,\sigma,
\end{aligned} \qquad (2.213)$$

where we have defined new coefficients $A, \dots, c$. Collecting terms in (2.213) according to the definition (2.206), the new coefficients are

$$A = \int_{z_O}^{z_I} (L\,g^4 + M\,g^2\,g'^2 + N\,g'^4)\,dz$$

$$B = \int_{z_O}^{z_I} (L\,h^4 + M\,h^2\,h'^2 + N\,h'^4)\,dz$$

$$C = \int_{z_O}^{z_I} (4\,L\,g^2 h^2 + 4\,M\,g\,h\,g'\,h' + 4\,N\,g'^2\,h'^2 - 4\,K\,k^2\,\mathsf{p}^{-2})\,dz$$

$$\begin{aligned}
D = \int_{z_O}^{z_I} & [\,2\,L\,g^2\,h^2 + M\,(g^2\,h'^2 + h^2\,g'^2) + 2\,N\,g'^2\,h'^2 \\
& +4\,K\,k^2\,\mathsf{p}^{-2}\,]\,dz
\end{aligned}$$

$$E = \int_{z_O}^{z_I} [4\,L\,g^3\,h + 2\,M\,g\,g'\,(g\,h' + g'\,h) + 4\,N\,g'^3\,h']\,dz$$

$$F = \int_{z_O}^{z_I} [4\,L\,g\,h^3 + 2\,M\,h\,h'\,(g\,h' + g'\,h) + 4\,N\,g'\,h'^3]\,dz$$

$$e = 2\,k \int_{z_O}^{z_I} \mathsf{p}^{-1}\,(P\,g^2 + Q\,g'^2)\,dz$$

$$c = 4\,k \int_{z_O}^{z_I} \mathsf{p}^{-1}\,(P\,g\,h + Q\,g'\,h')\,dz$$

$$f = 2\,k \int_{z_O}^{z_I} \mathsf{p}^{-1}\,(P\,h^2 + Q\,h'^2)\,dz, \tag{2.214}$$

remembering that $g(z)$ and $h(z)$ are the two linearly independent solutions to the paraxial ray equation (2.162) satisfying boundary conditions (2.166). From (2.213) we have

$$\frac{\partial}{\partial x_{Aj}} \int_{z_O}^{z_I} m_4\,dz = A\,\frac{\partial}{\partial x_{Aj}}\,(R^2) + B\,\frac{\partial}{\partial x_{Aj}}\,(\rho^2) + C\,\frac{\partial}{\partial x_{Aj}}\,(\chi^2) + \dots \tag{2.215}$$

The primary aberration δx_I in the Gaussian image plane is thus given in the rotated coordinate system by (2.203, 2.215)

$$
\begin{aligned}
k\,\mathrm{M}^{-1}\,\delta x_I = \quad & B\,[\,4\,x_A\,(x_A^2 + y_A^2)\,] \\
+\ & C\,[\,2\,x_O\,(x_O x_A + y_O y_A)\,] \\
+\ & D\,[\,2\,x_A\,(x_O^2 + y_O^2)\,] \\
+\ & E\,[\,x_O\,(x_O^2 + y_O^2)\,] \\
+\ & F\,[\,x_O\,(x_A^2 + y_A^2) + 2\,x_A\,(x_O x_A + y_O y_A)\,] \\
+\ & e\,[\,-y_O\,(x_O^2 + y_O^2)\,] \\
+\ & c\,[\,y_A\,(x_O^2 - y_O^2) - 2\,x_O\,y_O\,x_A\,] \\
+\ & f\,[\,-y_O\,(x_A^2 + y_A^2) + 2\,x_A\,(x_O y_A - y_O x_A)\,],
\end{aligned}
\tag{2.216}
$$

remembering the definition (2.165) of the constant k, and where M is the magnification. Making use of the axial symmetry, the y-coordinate of the aberration is obtained by making the substitutions $x \to y$, and $y \to -x$ for all occurrences in (2.216).

This gives

$$
\begin{aligned}
k\,\mathrm{M}^{-1}\,\delta y_I = \quad & B\left[4\,y_A\left(x_A^2 + y_A^2\right)\right] \\
+ \; & C\left[2\,y_O\left(x_O x_A + y_O y_A\right)\right] \\
+ \; & D\left[2\,y_A\left(x_O^2 + y_O^2\right)\right] \\
+ \; & E\left[y_O\left(x_O^2 + y_O^2\right)\right] \\
+ \; & F\left[y_O\left(x_A^2 + y_A^2\right) + 2\,y_A\left(x_O x_A + y_O y_A\right)\right] \\
+ \; & e\left[x_O\left(x_O^2 + y_O^2\right)\right] \\
+ \; & c\left[-x_A\left(x_O^2 - y_O^2\right) + 2\,x_O\,y_O\,y_A\right] \\
+ \; & f\left[x_O\left(x_A^2 + y_A^2\right) + 2\,y_A\left(x_O y_A - y_O x_A\right)\right].
\end{aligned}
\tag{2.217}
$$

We notice that

$$
\delta x_{Oj} = \mathrm{M}^{-1}\,\delta x_{Ij} \tag{2.218}
$$

is the aberration, demagnified to the object plane z_O. We call δx_{Oj} the aberration *referred* to the object plane. This is of interest for a transmission electron microscope, for example, where the object coordinates form the natural reference for the expression of image quality. Similarly, one has the option of substituting $x_O = \mathrm{M}^{-1} x_I$ and $y_O = \mathrm{M}^{-1} y_I$ on the right sides of (2.216, 2.217), thus referring the aberrations to the image plane z_I. This is of interest for a probe forming system, like a scanning electron microscope or focused ion beam system, where one typically forms a demagnified image of a source. In this case, the image coordinates form the natural reference. Either object or image coordinates correctly express the aberrations.

A significant simplification is possible by choosing the rotation of coordinate axes so that $y_O = 0$; i.e., the off-axis object position is located along the x-axis. There is no loss of generality, owing to the axial symmetry, as the coordinates for any *single* object point can always be chosen in this way.

In this case the primary aberration $(\delta x_I, \delta y_I)$ simplifies (2.216, 2.217) to

$$
\begin{aligned}
k\,\mathrm{M}^{-1}\,\delta x_I \;=\;& 4\,B\,x_A\,(x_A^2 + y_A^2) + 2\,(C+D)\,x_O^2\,x_A + E\,x_O^3 \\
& + F\,x_O\,(\,3\,x_A^2 + y_A^2\,) + c\,x_O^2\,y_A + 2\,f\,x_O\,x_A\,y_A
\end{aligned}
$$

$$
\begin{aligned}
k\,\mathrm{M}^{-1}\,\delta y_I \;=\;& 4\,B\,y_A\,(x_A^2 + y_A^2) + 2\,D\,x_O^2\,y_A + 2\,F\,x_O\,x_A\,y_A \\
& + e\,x_O^3 + c\,x_O^2\,x_A + f\,x_O\,(x_A^2 + 3\,y_A^2).
\end{aligned}
$$

$$(2.219)$$

The various series representations of the aberration are known as a *Seidel* series. The individual terms in (B, C, D, E, F, e, c, f) represent, respectively, spherical aberration, isotropic astigmatism, field curvature, isotropic distortion, isotropic coma, anisotropic distortion, anisotropic astigmatism, and anisotropic coma. They are referred to as third order aberrations, because each term is third order in various products including (x_O, y_O, x_A, y_A). This represents the solution for the primary aberrations. All quantities representing length have the same values in natural units and SI units. This includes the aberrations δx_I and δy_I. However, the axial potential $\Phi(z)$ and axial magnetic field $\mathrm{B}(z)$, which form the basis of the field coefficients (2.206), do depend on the choice of units.

The preceding results give the aberration for a single ray. In practice, a beam is comprised of a bundle of rays, each having a different aberration δx_{Ij} in the Gaussian image plane. Even in the limit of an ideal point object, these aberrations cause blurring of the image. The amount of blurring varies with defocus. It is therefore of interest to study the aberration in a plane which is slightly displaced from the Gaussian image plane. Designating the axial displacement by δz, we seek an expression for the aberration δx_j in the plane $z_I + \delta z$. This is given to first order in δz by

$$
\begin{aligned}
x_{Ij} + \delta x_j &= x_{Ij} + \delta x_{Ij} + x'_{Ij}\,\delta z \\
\delta x_j &= \delta x_{Ij} + x'_{Ij}\,\delta z, \qquad (2.220)
\end{aligned}
$$

where x_{Ij} is the paraxial ray coordinate in the Gaussian image plane, δx_{Ij} is the primary aberration in the Gaussian image plane, and x'_{Ij} is the paraxial ray slope in the Gaussian image plane. Evaluating (2.168) in the Gaussian image plane, this gives

$$x'_{Ij} = x_{Oj}\, g'_I + x_{Aj}\, h'_I. \tag{2.221}$$

The aberration δx_j with defocus δz is thus given by

$$\delta x_j = \delta x_{Ij} + (x_{Oj}\, g'_I + x_{Aj}\, h'_I)\, \delta z. \tag{2.222}$$

For a fixed value of δz, this is computed separately for every ray in the bundle, which in principle enables the blur to be found as a function of defocus δz.

We define a quantity

$$W_4 = \int_{z_O}^{z_I} m_4\, dz. \tag{2.223}$$

We recognize $W_4 = \delta W_2$ as the first order perturbation on the action integral (2.194) which gives rise to the primary aberration. All rays emanating from any single object point have the same value of W_2, corresponding to the paraxial approximation. Each of these rays has a unique value of W_4 which in general differs from the other rays, corresponding to the aberration. The above analysis shows that all relevant information about the primary aberration is contained in W_4.

One can derive higher order aberrations by considering the terms $m_6, m_8, \ldots$ in (2.159). These aberrations are referred to as fifth order, seventh order, $\ldots$, respectively. They have corresponding perturbations $W_6, W_8, \ldots$ in the action integral. This procedure is straightforward, but tedious, since each increasing order contains more terms than the preceding one. The number of terms needed to obtain an accurate representation depends on the size of the ray coordinates $x_j(z)$ and slopes $x'_j(z)$. This in turn depends on the lateral extent of the beam, as determined by the physical aperture. A narrow beam requires fewer terms than a wide beam. The exact value of the action integral between the object and image planes is given by

$$W_{OI} = \int_{z_O}^{z_I} m(z)\, dz, \tag{2.224}$$

where $m(z)$ is the sum of all terms in the infinite series (2.159). In principle, all relevant information about the optical system is contained in W in the limit of geometrical optics. This fact will prove to be highly useful in the analysis to come.

2.5.6 Spherical aberration

In the case where the object is on the optic axis, we have $x_{Oj} = 0$. In this case, the primary aberration in the Gaussian image plane reduces to (2.216, 2.217)

$$\delta x_{Ij} = \frac{4\,B\,M}{k}\,(x_A^2 + y_A^2)\,x_{Aj}. \tag{2.225}$$

All of the aberrations vanish on axis except spherical aberration, represented by the B-term. Spherical aberration is the same everywhere in the field, as it is independent of object coordinate x_{Oj}. It only depends on the coordinate x_{Aj} in the aperture plane. Spherical aberration is shown schematically in Figure 2.8. The axis of symmetry is the central ray in the bundle. Rays close to this axis

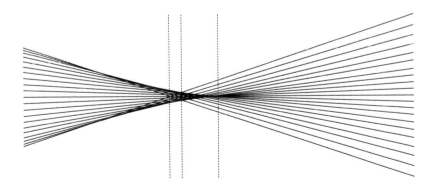

Figure 2.8: Spherical aberration.

come to a focus at the Gaussian image plane, which is shown by
the rightmost vertical line in the figure. Rays at the edge are more
strongly focused, and intersect the axis in front of the Gaussian
image plane. This plane is shown by the leftmost vertical line in
the figure. A disk of least confusion is formed by the entire bundle
at an intermediate plane. This represents the optimum focus.

Spherical aberration is easily expressed as a function of the ray
slope x'_{Ij} in the Gaussian image plane. This slope is directly measurable by defocusing, whereas the aperture coordinate x_{Aj} is
not easily measured. This necessitates transforming from the set
(x_{Oj}, x_{Aj}) to the set (x_{Oj}, x'_{Ij}). Evaluating (2.168) in the Gaussian
image plane, and setting $x_{Oj} = 0$, we find

$$x_{Aj} = \frac{x'_{Ij}}{h'_I}. \tag{2.226}$$

From (2.169) we have

$$\frac{1}{h'_I} = \frac{\mathsf{p}_I\, \mathrm{M}}{k}, \tag{2.227}$$

in which case

$$x_{Aj} = \frac{\mathsf{p}_I\, \mathrm{M}}{k}\, x'_{Ij}. \tag{2.228}$$

We now define radial quantities r_A, r'_I, δr_I as

$$
\begin{aligned}
r_A^2 &= x_A^2 + y_A^2 \\
r'^2_I &= x'^2_I + y'^2_I \\
\delta r_I^2 &= \delta x_I^2 + \delta y_I^2,
\end{aligned}
\tag{2.229}
$$

from which, due to the axial symmetry, it follows that

$$\delta r_I = \frac{4\,B\,\mathrm{M}}{k}\, r_A^3 = \frac{4\,B\,\mathrm{M}^4\, \mathsf{p}_I^3}{k^4}\, r'^3_I. \tag{2.230}$$

We define α_I as the angle in radians which the ray makes with the
optic axis at the Gaussian image plane. Assuming $\alpha \ll 1$,

$$r'_I = \tan \alpha_I \approx \alpha_I. \tag{2.231}$$

This enables us to write, in this small angle approximation,

$$\delta r_I = C_{SI}\, \alpha_I^3, \qquad (2.232)$$

where we have defined a constant C_{SI}, called the coefficient of spherical aberration, referred to the image, as

$$C_{SI} = \frac{4\,B\,M^4\,\mathsf{p}_I^3}{k^4}, \qquad (2.233)$$

where C_{SI} depends on the axial electrostatic potential $\Phi(z)$ and axial magnetic field $\mathsf{B}(z)$ through the coefficient B in (2.214).

Alternatively, spherical aberration can be referred to the object plane by making use of

$$\delta r_I = \mathrm{M}\,\delta r_O, \qquad (2.234)$$

where δr_O is the aberration in the object plane z_O. Applying the law of Helmholtz-Lagrange (2.69), we have

$$\mathsf{p}_O\,\alpha_O\,(\delta r_O) = \mathsf{p}_I\,\alpha_I\,(\delta r_I), \qquad (2.235)$$

relating the object and image planes. It follows that the angles in the object and image planes are related by

$$\alpha_I = \left(\frac{\mathsf{p}_O}{\mathsf{p}_I\,\mathrm{M}}\right)\alpha_O, \qquad (2.236)$$

where the parenthesis on the right is the angular magnification. The spherical aberration in the object plane is then

$$\delta r_O = C_{SO}\,\alpha_O^3, \qquad (2.237)$$

where we have defined the spherical aberration coefficient C_{SO}, referred to the object, as

$$C_{SO} = \left(\frac{\mathsf{p}_O}{\mathsf{p}_I}\right)^3 \frac{C_{SI}}{\mathrm{M}^4}. \qquad (2.238)$$

It is natural to refer the spherical aberration to the object plane in a transmission electron microscope, and to the image plane in probe-forming systems such as a scanning electron microscope.

It was first shown by Scherzer [77] that spherical aberration cannot
be eliminated in static systems with axial symmetry in the absence
of space charge, and excluding particle mirrors. This is equivalent
to the coefficient B defined in (2.214) always being positive definite. This fact can be proven by successive partial integrations of
the expression for B, in which the integrand can be expressed by a
sum of positive definite terms. The reader is referred to Rose [75]
for details.

Spherical aberration imposes a fundamental limit on the resolution of electron microscopes. Substantial correction of spherical
aberration has been successfully demonstrated using multipole elements [67, 75]. Corrected electron microscopes are now commercially available which achieve resolution better than 0.1 nm. This
is sufficient to view single atoms.

2.5.7   Field aberrations

The aberrations represented by the terms proportional to C, D, E,
F, e, c, f in (2.216, 2.217) represent isotropic astigmatism, field
curvature, isotropic distortion, isotropic coma, anisotropic distortion, anisotropic astigmatism, and anisotropic coma, respectively.
Unlike spherical aberration, all of these aberrations depend on
field position, as represented by the object coordinates (x_O, y_O).
We therefore call them *field* aberrations. Isotropic aberrations
are characterized by having no dependence on azimuthal coordinate, while anisotropic aberrations have an azimuthal dependence.
The isotropic aberrations arise in both electrostatic and magnetic
lenses, while anisotropic aberrations only occur in magnetic lenses.
The field aberrations can be best appreciated by simply plotting
the geometric figure, which in general is a function of the object
coordinates (x_O, y_O) and the aperture coordinates (x_A, y_A). In this

section we discuss the field aberrations individually.

The aberrations represented by the terms proportional to C and c in (2.216, 2.217) represent isotropic astigmatism and anisotropic astigmatism, respectively. The aberration figure for astigmatism is plotted in Figure 2.9. The beam forms two line foci, separated by

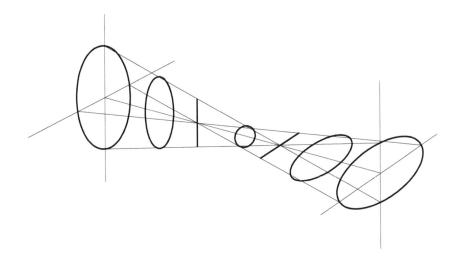

Figure 2.9: Astigmatism.

an axial distance, and oriented at 90 degrees relative to each other. The axial separation of the line foci is proportional to the square of the radial distance off axis of the object point. The primary astigmatism aberration vanishes for an object point on axis. The line foci are oriented along the x- and y-axes for isotropic astigmatism, and at 45 degrees to these axes for anisotropic astigmatism. The beam cross section is axially symmetric at an axial point midway between the two line foci. By adjusting the focus, one is able to locate this plane, which is characterized by an image which is axially symmetric, though blurred. Alternatively, astigmatism can arise from misalignment of the optical elements, or departure from axial symmetry. Although the manifestation appears similar, the mechanism by which the astigmatism arises is quite different from the primary aberration discussed here.

The aberration represented by the term proportional to D in (2.216, 2.217) represents curvature of field. This aberration results in an axial displacement of the plane of best focus which is proportional to the square of the radial distance off axis. The aberration figure for curvature of field is plotted in Figure 2.10. The figure

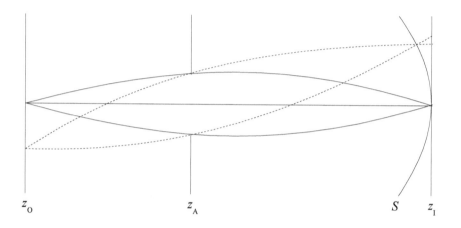

Figure 2.10: Curvature of field.

schematically depicts a longitudinal section through the optical system. The solid curved rays represent an object point on axis, for which the image plane z_I coincides with the Gaussian image plane. The broken curved rays represent the aberrated rays for an object point off axis, for which the plane of best focus lies on the curve S. The focal surface for off-axis object points can be generated in three dimensions by rotating the curve labeled S about the central optic axis. This surface therefore has axial symmetry.

The aberrations represented by the terms proportional to E and e in (2.216, 2.217) represent isotropic distortion and anisotropic distortion, respectively. The aberration figures for positive and negative isotropic distortion are plotted in Figure 2.11. Isotropic distortion results in a radial displacement of the image point by an amount which is proportional to the cube of the radial distance off axis. The aberration figures for positive and negative anisotropic

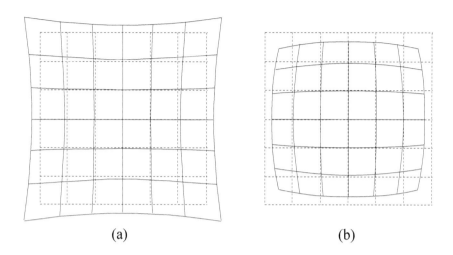

(a)   (b)

Figure 2.11: Isotropic distortion, (a) pincushion, (b) barrel.

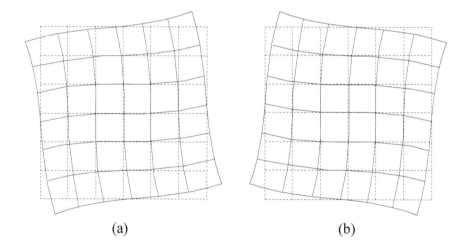

(a)   (b)

Figure 2.12: Anisotropic distortion, (a) positive, (b) negative.

distortion are plotted in Figure 2.12. Anisotropic distortion results in an azimuthal displacement of the image point by an amount which is proportional to the cube of the radial distance off axis.

The aberrations represented by the terms proportional to F and f

in (2.216, 2.217) represent isotropic coma and anisotropic coma, respectively. The aberration figure for coma is plotted in Figure 2.13. Each circle represents the aberrated transverse position in the

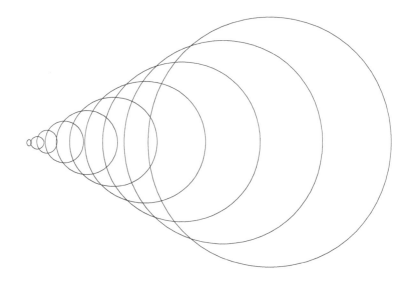

Figure 2.13: Coma.

Gaussian image plane for a given radial coordinate in the aperture plane. The smallest circle corresponds to the zone nearest to the center of the aperture, while the largest circle corresponds to the zone nearest to the rim of the aperture. The figure with all of the circles is plotted for a single point in the object plane. Physically, coma arises because the transverse magnification varies with radial coordinate in the aperture plane. Isotropic coma has a figure which is aligned with the radial direction in the Gaussian image plane, while anisotropic coma has a figure which is oriented in the azimuthal direction. The intensity represnts a blur which resembles a comet, hence the name coma. One can show that the angle between the two lines which form the envelope for all of the circles is always 60 degrees.

2.5.8 Chromatic aberration

We now turn our attention to the aberration which arises when an individual particle has a kinetic energy which differs infinitesimally by $\delta\Phi$ from the nominal kinetic energy on axis $\Phi(z)$. This arises quite commonly, as every practical particle source emits with a range or spread of kinetic energies at the emission surface. As the focussing action of electric and magnetic fields depends on the particle energy, we expect an aberration, called the *chromatic* aberration to occur. This name originates from the analogy between particle optics and light optics, where the color or chromaticity of the light is directly related to the photon energy. The following analysis is based on that of Zworykin, et. al. [94].

The problem can be stated mathematically as follows: given the solution $v(z)$ to the paraxial ray equation for energy $\Phi(z)$ on axis, find the aberration δv_I in the Gaussian image plane, arising from a constant perturbation $\delta\Phi$ in the energy. We begin by recalling that $v(z)$ is the general solution (2.167) to the paraxial ray equation (2.162) in the rotated system. Expanding the first derivative in (2.162), we obtain the equivalent paraxial ray equation

$$v''(z) + \left[\mathsf{p}^{-2}\, \Phi'\, (1 + \Phi) \right] v'(z)$$
$$+ \left[\tfrac{1}{2}\, \mathsf{p}^{-2}\, \Phi''\, (1 + \Phi) + \tfrac{1}{4}\, \mathsf{p}^{-2}\, \mathsf{B}^2 \right] v(z) = 0. \quad (2.239)$$

We define the perturbed ray and energy as

$$
\begin{aligned}
v_1(z) &= v(z) + \delta v_1(z) \\
\Phi_1(z) &= \Phi(z) + \delta\Phi,
\end{aligned}
\qquad (2.240)
$$

respectively. We assume $\delta\Phi = const.$

We know a priori that $v_1(z)$ must be a solution to the paraxial ray equation (2.239) with energy $\Phi_1(z)$. Substituting (2.240) into (2.239), and canceling the unperturbed terms, it is tedious, but straightforward to show that $\delta v_1(z)$ satisfies

$$
\begin{aligned}
(\delta v_1)'' \;+\;& \left[\, \mathsf{p}^{-2}\,\Phi'\,(1+\Phi)\right] (\delta v_1)' \\
+\;& \left[\tfrac{1}{2}\mathsf{p}^{-2}\,\Phi''\,(1+\Phi) + \tfrac{1}{4}\mathsf{p}^{-2}\,\mathsf{B}^2\right] (\delta v_1) \\
=\;& (\delta\Phi)\,\mathsf{p}^{-2}\,\{\,(2\,\mathsf{p}^{-2}+1)\,\Phi'\,v' \\
+\;& \tfrac{1}{2}\left[\,(2\,\mathsf{p}^{-2}+1)\,\Phi'' + \mathsf{p}^{-2}(1+\Phi)\,\mathsf{B}^2\,\right] v\,\},\quad (2.241)
\end{aligned}
$$

retaining only terms to first order in δv_1 and $\delta\Phi$. This is an in-homogeneous second order differential equation in δv_1, due to the nonzero right-hand side.

The general solution to such an inhomogeneous equation can al-ways be expressed as the sum of the solution to the homogeneous equation, plus any particular solution to the inhomogeneous equa-tion. The left side is identical with the left side of the paraxial ray equation (2.239), with $v(z)$ replaced by $\delta v_1(z)$. The homoge-neous solution is therefore (2.167), with $v(z)$ replaced by $\delta v_1(z)$. The independent solutions $g(z)$ and $h(z)$ are replaced by perturbed solutions, designated by $g+\delta g$ and $h+\delta h$, respectively. The pertur-bations δg and δh do not appear in the first order approximation (2.241), however, and can be ignored.

The general solution for $\delta v_1(z_I)$, evaluated in the Gaussian im-age plane of the unperturbed ray $v(z)$ is

$$
\delta v_1(z_I) = -\frac{\mathrm{M}}{k}\int_{z_O}^{z_I} \mathsf{p}(z)\, S(z)\, h(z)\, dz, \qquad (2.242)
$$

where M is the magnification, k is the conserved Wronskian (2.165, 2.169), and $S(z)$ is the right-hand side of (2.241), namely,

$$
\begin{aligned}
S(z) \;=\;& (\delta\Phi)\,\mathsf{p}^{-2}\,\{\,(2\,\mathsf{p}^{-2}+1)\,\Phi'\,v' \\
&+\tfrac{1}{2}\left[\,(2\,\mathsf{p}^{-2}+1)\,\Phi'' + \mathsf{p}^{-2}(1+\Phi)\,\mathsf{B}^2\,\right] v\,\}.\quad (2.243)
\end{aligned}
$$

The solution (2.242) is derived in Appendix B, along with the gen-eral method of solving an inhomogeneous second order differential equation.

Having solved for the perturbation $\delta v_1(z_I)$, we must now express this in the unperturbed coordinate system $v(z)$. Because the ray $v_1(z)$ has energy Φ_1 which differs infinitesimally from Φ, it follows that the rotation $\chi(z)$ differs by an infinitesimal amount $\delta\chi$ between the v- and v_1-systems. Applying this rotation, we define the perturbation $\delta v(z_I)$ in the *unperturbed* coordinates $v(z)$ according to

$$
\begin{aligned}
v + \delta v &= v_1\, e^{-i\,\delta\chi} \\
&\approx (v + \delta v_1)\,(1 - i\,v\,\delta\chi).
\end{aligned}
\tag{2.244}
$$

Expanding this, the aberration expressed in the Gaussian image plane z_I of the unperturbed system is

$$
\delta v(z_I) = \delta v_1(z_I) - i\,v(z_I)\,\delta\chi_{OI},
\tag{2.245}
$$

retaining only terms to first order in small quantities. From (2.157), the perturbation $\delta\chi$ is found to be

$$
\delta\chi_{OI} = \tfrac{1}{2}\int_{z_O}^{z_I}\delta(\mathsf{p}^{-1})\,\mathsf{B}\,dz = -(\delta\Phi)\cdot\tfrac{1}{2}\int_{z_O}^{z_I}\mathsf{p}^{-3}(1+\Phi)\,\mathsf{B}\,dz,
\tag{2.246}
$$

where we have made use of (2.124), the definition of the on-axis kinetic momentum p. The minus sign expresses the fact that the rotation χ is smaller for higher particle energy, $\delta\Phi > 0$. Substituting (2.242, 2.243, 2.246) into (2.245), and making use of the solution (2.167), we find, after collecting terms,

$$
\delta v_I = (\delta\Phi)\,[\,(C_1 + iC_2)\,v_O + C_3\,v_A\,],
\tag{2.247}
$$

where we have defined the chromatic aberration coefficients C_1, C_2, and C_3 in natural units as

$$
\begin{aligned}
C_1 &= -\frac{\mathsf{M}}{k}\int_{z_O}^{z_I}\mathsf{p}^{-1}\big\{\,(2\,\mathsf{p}^{-2}+1)\,\Phi'\,g'\,h \\
&\qquad + \tfrac{1}{2}\big[\,(2\,\mathsf{p}^{-2}+1)\,\Phi'' + \mathsf{p}^{-2}(1+\Phi)\,\mathsf{B}^2\,\big]\,g\,h\,\big\}\,dz \\[6pt]
C_2 &= \frac{\mathsf{M}}{2}\int_{z_O}^{z_I}\mathsf{p}^{-3}\,(1+\Phi)\,\mathsf{B}\,dz \\[6pt]
C_3 &= -\frac{\mathsf{M}}{k}\int_{z_O}^{z_I}\mathsf{p}^{-1}\big\{\,(2\,\mathsf{p}^{-2}+1)\,\Phi'\,h\,h' \\
&\qquad + \tfrac{1}{2}\big[\,(2\,\mathsf{p}^{-2}+1)\,\Phi'' + \mathsf{p}^{-2}(1+\Phi)\,\mathsf{B}^2\,\big]\,h^2\,\big\}\,dz.
\end{aligned}
\tag{2.248}
$$

By inspection, the three terms in (2.247) represent, in order, field magnification, field rotation, and defocus. This last is independent of field position. The chromatic aberration (2.247) is referred to the object by substituting $\delta v_O = \mathrm{M}^{-1} \delta v_I$. In the non-relativistic limit we substitute in (2.248) to obtain

$$1 + \Phi \approx 1, \qquad \mathsf{p} \approx \sqrt{2\,\Phi}, \qquad 2\,\mathsf{p}^{-2} + 1 \approx \frac{1}{\Phi}. \qquad (2.249)$$

The chromatic aberration δv_I can be expressed in Cartesian coordinates in the rotated system by substituting $v = x + i\,y$. Upon separating the real and imaginary terms, this gives (2.247)

$$\begin{aligned} \delta x_I &= (\delta\Phi)\,(C_1\,x_O - C_2\,y_O + C_3\,x_A) \\ \delta y_I &= (\delta\Phi)\,(C_1\,y_O + C_2\,x_O + C_3\,y_A). \end{aligned} \qquad (2.250)$$

This represents the solution for the chromatic aberration in the rotated coordinate system. The reader is reminded that δx_I and δy_I are quantitatively identical in natural units and in SI units, since the dimension of length is the same in both sets of units.

2.5.9 Intensity point spread function

The net effect of geometrical aberrations and defocus is that, even in the limit of a hypothetical ideal point object, the image is not a point, but is blurred. The amount of blurring gives a direct estimate of the quality of the image. In classical geometrical optics, all relevant information about the image is contained in the intensity as a function of position in the transverse plane. The physical image intensity can be regarded as a two-dimensional convolution of the ideal image intensity with an intensity point spread function. The intensity point spread function is the image of a hypothetical ideal point object, in the presence of aberrations and defocus. A method for calculating the intensity point spread function is derived by Gallatin [32]. In this section, we give an alternative derivation which is consistent with the foregoing analysis, and leads to the same result obtained by Gallatin.

In mathematical terms, we define the intensity point spread function $I(\mathbf{x}_I)$, as the two-dimensional intensity distribution in the Gaussian image plane for an ideal point object, in the classical limit of geometrical optics. We wish to obtain an analytic expression for $I(\mathbf{x}_I)$, given the aberrations and defocus. We can write

$$I(\mathbf{x}_I) = \int d^2\mathbf{P}_I \, \rho_I(\mathbf{x}_I, \mathbf{P}_I), \qquad (2.251)$$

where $\mathbf{x}_I = (x_I, y_I)$ is the transverse position, in the rotated system, $\mathbf{P}_I = (P_{Ix}, P_{Iy})$ is the transverse canonical momentum, and $\rho_I(\mathbf{x}_I, \mathbf{P}_I)$ is the phase space density, all defined in the Gaussian image plane. We have integrated over all momentum components, to obtain the intensity as a function of transverse position only.

Any optical system, however complicated, can be analyzed in terms of an equivalent system consisting of two lenses. This is shown schematically in Figure 2.14. In the equivalent system the object plane coincides with the front focal plane of the first lens.

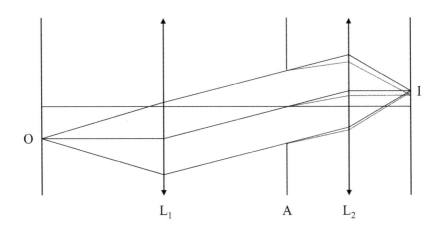

Figure 2.14: Equivalent confocal system.

A physical aperture is located at the back focal plane of the first lens, which coincides with the front focal plane of the second lens. It is easy to verify from the figure that the Gaussian image plane coincides with the back focal plane of the second lens. Both lenses are assumed to be ideal in the equivalent system. Each ray emanating from a point object intersects the aperture plane at a unique transverse position $\mathbf{x}_A = (x_A, y_A)$. Because the lenses are assumed to be perfect, every ray has the same optical path length between object and image. In the real system, the various rays have differing optical path length, owing to the aberrations.

The optical path difference for the primary aberration is given by (2.223, 2.159)

$$W_4 = \int_{z_O}^{z_I} m_4 \, dz \qquad (2.252)$$

for the real system. We assume this to be known for each ray from the preceding analysis. We now assume that all aberration of the *real* system for a particular ray is concentrated in the aperture plane of the *equivalent* system, manifest as an optical path difference W_4 from ideal.

In mathematical terms, we wish to find the intensity point spread function (2.251) in the Gaussian image plane, given the optical path difference (2.252) for each ray in the real system. For the equivalent system we can express the momentum element in the Gaussian image plane in terms of a unique area element in the aperture plane as

$$d^2\mathbf{P}_I = \left(\frac{\partial P_{Ix}}{\partial x_A} \frac{\partial P_{Iy}}{\partial y_A} - \frac{\partial P_{Ix}}{\partial y_A} \frac{\partial P_{Iy}}{\partial x_A} \right) d^2\mathbf{x}_A = \left(\frac{\mathsf{p}_I}{f} \right)^2 d^2\mathbf{x}_A, \quad (2.253)$$

where the large parenthesis is the Jacobian determinant, and where we have made use of (2.196)

$$\begin{aligned} \mathbf{P}(z) &= \mathsf{p}(z) \left[\mathbf{x}_O \, g'(z) + \mathbf{x}_A \, h'(z) \right] \\ \mathbf{P}_I &= \mathsf{p}_I \left(\mathbf{x}_O \, g_I' + \mathbf{x}_A \, h_I' \right) \end{aligned} \qquad (2.254)$$

in the paraxial approximation, where $\mathsf{p}(z)$ is the scalar kinetic momentum on axis. Also, $h_I' = 1/f$ where f is focal length of the final lens in the ideal system.

From Liouville's theorem we can relate the phase space density in the aperture and image planes of the equivalent system as follows:

$$\rho_I (\mathbf{x}_I, \mathbf{P}_I) = \rho_A (\mathbf{x}_A, \mathbf{P}_A). \qquad (2.255)$$

It follows that the intensity in the image plane can be written as (2.251, 2.253, 2.255)

$$I(\mathbf{x}_I) = \left(\frac{\mathsf{p}_I}{f}\right)^2 \int d^2\mathbf{x}_A \, \rho_A(\mathbf{x}_A, \mathbf{P}_A). \qquad (2.256)$$

This expresses the intensity in the Gaussian image plane entirely in terms of quantities in the aperture plane of the equivalent system. The reason for choosing the equivalent system becomes clear from this. We can assume a simple form for ρ_A as follows:

$$\rho_A (\mathbf{x}_A, \mathbf{P}_A) = T(\mathbf{x}_A) \, \delta(\mathbf{P}_A - \nabla W_A), \qquad (2.257)$$

where $T(\mathbf{x}_A)$ is assumed to be uniform over the aperture, corresponding to uniform illumination. Normalizing the area integral to unity, we set

$$T(\mathbf{x}_A) = 1/A \qquad (2.258)$$

inside the aperture, where we define A as the area of the aperture in the equivalent system.

The momentum distribution in (2.257) is a Dirac delta function, where ∇W_A is the two-dimensional gradient in the aperture plane (2.60). From (2.109) this is the transverse canonical momentum in the aperture plane. In the limit of perfect imaging, the surfaces of constant optical path are planar in the space between the two lenses of the equivalent system. This corresponds to a parallel beam of rays originating from a single object point. The intersection of these surfaces with the aperture plane form straight lines (for a general off-axis object point), which represent the contours of $W_A = const$. In the general case with aberrations, these contours are curved, owing to the gradient of the optical path difference across the aperture plane of the equivalent system. From (2.196)

$$\mathbf{P}_A = \mathsf{p}_A\left(\mathbf{x}_O\, g'_A + \mathbf{x}_A\, h'_A\right) = \mathsf{p}_A\, \mathbf{x}_I/f, \qquad (2.259)$$

where we have made use of

$$\mathbf{x}_O = \frac{\mathbf{x}_I}{M}, \qquad g'_A = \frac{M}{f}, \qquad h'_A = 0. \qquad (2.260)$$

We also assume the equivalent system to be monoenergetic in the space between the aprture plane z_A and the image plane z_I. It follows that $\mathsf{p}_A = \mathsf{p}_I \equiv \mathsf{p}$. From (2.256, 2.257, 2.258, 2.259)

$$I(\mathbf{x}_I) = \frac{1}{A}\int dx_A \int dy_A\, \delta\left(\frac{f}{\mathsf{p}}\frac{\partial W_A}{\partial x_A} - x_I\right)\delta\left(\frac{f}{\mathsf{p}}\frac{\partial W_A}{\partial y_A} - y_I\right), \qquad (2.261)$$

where we have made use of the properties of the delta function:

$$\delta(-x) = \delta(x), \qquad \delta(ax) = \frac{1}{a}\delta(x). \qquad (2.262)$$

Applying the property of delta function, we first define a function

$$f(a, b) = \int dx \int dy\, \delta[u(x, y) - a]\, \delta[v(x, y) - b]. \qquad (2.263)$$

In order to evaluate this, we must transform from the set (x, y) to the set (u, v). This involves the Jacobian determinant,

$$du\, dv = \left(\frac{\partial u}{\partial x}\frac{\partial v}{\partial y} - \frac{\partial u}{\partial y}\frac{\partial v}{\partial x}\right) dx\, dy \equiv D(u, v)\, dx\, dy. \qquad (2.264)$$

With the coordinate transformation complete, the integral is evaluated using the property of the delta function,

$$f(a, b) = \int du \int dv\, \frac{1}{D(u, v)}\, \delta(u - a)\, \delta(v - b) = \frac{1}{D(a, b)}. \qquad (2.265)$$

Applying this mathematical formalism to the present problem, we define (2.261)

$$u(x_A, y_A) = \frac{f}{\mathsf{p}}\frac{\partial}{\partial x_A} W_A(x_A, y_A)$$

$$v(x_A, y_A) = \frac{f}{\mathsf{p}}\frac{\partial}{\partial y_A} W_A(x_A, y_A). \qquad (2.266)$$

We assume $u(x_A, y_A)$ and $v(x_A, y_A)$ to be known functions, as $W_A(x_A, y_A)$ is known a priori. Substituting, we find

$$
\begin{aligned}
D(u, v) &= \left(\frac{\partial u}{\partial x_A} \frac{\partial v}{\partial y_A} - \frac{\partial u}{\partial y_A} \frac{\partial v}{\partial x_A} \right) \\
&= \left(\frac{f}{\rho} \right)^2 \left(\frac{\partial^2 W_A}{\partial x_A^2} \frac{\partial^2 W_A}{\partial y_A^2} - \frac{\partial^2 W_A}{\partial y_A \partial x_A} \frac{\partial^2 W_A}{\partial x_A \partial y_A} \right).
\end{aligned}
\tag{2.267}
$$

Consistent with the delta function (2.261), we set

$$
u(x_A, y_A) = x_I, \qquad v(x_A, y_A) = y_I, \tag{2.268}
$$

and invert this pair to solve for (x_A, y_A) in terms of (x_I, y_I). Since $W_A(x_A, y_A)$ is typically represented as a polynomial with terms $x_A^m \cdot y_A^n$, this inversion amounts to finding the roots of a polynomial.

Using this new solution for $(\tilde{x}_A, \tilde{y}_A)$, we form

$$
D[u(\tilde{x}_A, \tilde{y}_A), v(\tilde{x}_A, \tilde{y}_A)] \equiv D(x_I, y_I). \tag{2.269}
$$

The final result for intensity point spread function is then

$$
I(x_I, y_I) = \frac{1}{A \cdot D(x_I, y_I)}. \tag{2.270}
$$

Applying this procedure for any point in an extended object, one constructs the intensity point spread function about corresponding image point in the limit of geometrical optics. This is the main result of this section.

2.6 Stochastic Coulomb scattering

In an earlier section, we discussed the effect of space charge on a hypothetical test particle in the beam. The space charge was regarded as a continuum. Approximating the current density as uniform within the beam volume, we showed that the space charge acts as a negative lens. In the paraxial approximation, the net result of the space charge is defocusing. The space charge lens also has aberrations, which can be calculated in principle.

This approximation does not precisely agree with experiment, however. The first indication of this was seen by Boersch [7], who measured significant energy broadening in an electron beam, which grew monotonically with current. This could not be explained by any continuum approximation. This took on practical significance with the advent of electron beam lithography, for which the Coulomb interaction places a limit on the useful writing current for a given resolution. This in turn limits the throughput to values which are slow compared with optical lithography.

A beam of particles can be regarded more realistically as a collection of discrete, moving point charges, distributed randomly in space within the beam volume. Every particle exerts a Lorentz force (2.15) on every other particle, resulting in a random displacement of each particle. The effect becomes more pronounced as the beam current is increased, owing to the closer proximity of beam particles. It also becomes stronger as the interaction time is increased, as the particles have more time to interact. The interaction time increases as the length of the beam path increases, or the beam energy decreases.

A rough estimate of the relative strength of the interaction is obtained from the average axial spacing between particles, given by the charge times the velocity divided by the current. In a typical electron microscope, no more than one electron is in the column at any given instant on average. Coulomb scattering is unimportant in this case. In a typical electron or ion beam lithography system, one uses the highest possible current, without unacceptably degrading the resolution. Consequently, the axial distance between particles can be on the order of micrometers. In this case the interaction is important enough, that it imposes a limit on the useful writing current for a given resolution.

2.6.1 Monte Carlo simulation

Stochastic Coulomb scattering is a many-body interaction involving a large number of particles. As such, the detailed motion of every particle cannot be solved in closed form, even if the initial position and velocity of every particle were known. Fortunately, a great deal of understanding can be gained by treating the interaction statistically. Numerical Monte Carlo simulation [26, 39, 76] provides a tool for accurately predicting the relevant performance parameters for a given system configuration. A pseudo-random number generator is used to initialize the positions and velocities of many particles in the vicinity of the source, with the beam voltage and source current taken into account. The motion of every particle is then traced numerically through the optical system, with the Lorentz force due to every other particle taken into account at every step.

The Lorentz force is the vector sum of an electrostatic force and a magnetic force (2.15). This is shown schematically in Figure 2.15. A particle labeled a with charge q_a and mass m_a is at position $\mathbf{r}_a$ with velocity $\mathbf{v}_a$ in the lab frame. This particle experiences a Lorentz force due to a second particle labeled b with charge q_b and mass m_b at position $\mathbf{r}_b$ with velocity $\mathbf{v}_b$. The Lorentz force on particle a due to particle b is given in the lab frame by (2.15)

$$\frac{d}{dt}(\gamma_a m_a \mathbf{v}_a) = q_a(\mathbf{E}_{ab} + \mathbf{v}_a \times \mathbf{B}_{ab}), \qquad (2.271)$$

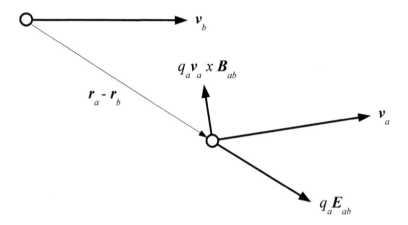

Figure 2.15: Lorentz force on a particle due to a second particle.

where $\mathbf{E}_{ab}$ is the electric field at particle a due to particle b, and $\mathbf{B}_{ab}$ is the magnetic field at particle a due to particle b. We can assume without loss of generality that the vectors $\mathbf{v}_b$ and $\mathbf{r}_a - \mathbf{r}_b$ determine a plane, and this plane coincides with the plane of the page in Figure 2.15. The electric field at the position of particle a due to particle b is given by

$$\mathbf{E}_{ab} = \frac{q_b}{4\pi\epsilon_0}\frac{\mathbf{r}_a - \mathbf{r}_b}{|\mathbf{r}_a - \mathbf{r}_b|^3}. \qquad (2.272)$$

The magnetic field is given by

$$\mathbf{B}_{ab} = \frac{\mu_0 q_b}{4\pi}\frac{\mathbf{v}_b \times (\mathbf{r}_a - \mathbf{r}_b)}{|\mathbf{r}_a - \mathbf{r}_b|^3}. \qquad (2.273)$$

This field points into the plane of the page for positive charge q_b. The Lorentz force on particle a is then

$$\frac{d}{dt}(\gamma_a m_a \mathbf{v}_a) = \frac{q_a q_b}{4\pi\epsilon_0 |\mathbf{r}_a - \mathbf{r}_b|^3}$$

$$\cdot \left\{ (\mathbf{r}_a - \mathbf{r}_b) + \frac{1}{c^2}\mathbf{v}_a \times [\mathbf{v}_b \times (\mathbf{r}_a - \mathbf{r}_b)] \right\},$$

$$(2.274)$$

where we have made use of $\mu_0\epsilon_0 = 1/c^2$. Expanding the double cross product, and summing over all particles b, the total vector Lorentz force on particle a due to all other particles is

$$\frac{d}{dt}(\gamma_a m_a \mathbf{v}_a) = \frac{q_a}{4\pi\epsilon_0} \sum_{b\neq a} \frac{q_b}{|\mathbf{r}_a - \mathbf{r}_b|^3}$$
$$\cdot \left\{ (\mathbf{r}_a - \mathbf{r}_b)\left(1 - \frac{\mathbf{v}_a \cdot \mathbf{v}_b}{c^2}\right) + \frac{1}{c^2}\mathbf{v}_b[\mathbf{v}_a \cdot (\mathbf{r}_a - \mathbf{r}_b)] \right\}.$$

$$(2.275)$$

To this point we have made no approximations.

We now approximate that $d\gamma_a/dt \approx 0$. The resultant acceleration $\mathbf{a}_a$ of particle a is then given by

$$\frac{d\mathbf{v}_a}{dt} = \frac{q_a}{4\pi\epsilon_0\gamma_a m_a} \sum_{b\neq a} \frac{q_b}{|\mathbf{r}_a - \mathbf{r}_b|^3}$$
$$\cdot \left\{ (\mathbf{r}_a - \mathbf{r}_b)\left(1 - \frac{\mathbf{v}_a \cdot \mathbf{v}_b}{c^2}\right) + \frac{1}{c^2}\mathbf{v}_b[\mathbf{v}_a \cdot (\mathbf{r}_a - \mathbf{r}_b)] \right\}.$$

$$(2.276)$$

We are now in a position to numerically compute the trajectory of particle a. Dropping the subscript a, we can write a Taylor series for the trajectory point $i + 1$ in terms of the point i as

$$\begin{aligned}
\mathbf{r}_{i+1} &= \mathbf{r}_i + \mathbf{v}_i(\Delta t) + \tfrac{1}{2}\mathbf{a}_i(\Delta t)^2 + \tfrac{1}{6}\dot{\mathbf{a}}_i(\Delta t)^3 + \dots \\
\mathbf{v}_{i+1} &= \mathbf{v}_i + \mathbf{a}_i(\Delta t) + \tfrac{1}{2}\dot{\mathbf{a}}_i(\Delta t)^2 + \dots \\
\mathbf{a}_{i+1} &= \mathbf{a}_i + \dot{\mathbf{a}}_i(\Delta t) + \dots,
\end{aligned} \qquad (2.277)$$

where the time increment is $\Delta t = t_{i+1} - t_i$. The quantity $\dot{\mathbf{a}}_i$ is the time rate of change of the acceleration $\mathbf{a}_i$. This can be calculated analytically by time differentiation of the expression for the acceleration. This is left as an exercise for the reader. This procedure is repeated for all of the particles in the sample.

The physical significance of the interaction can be better appreciated by noticing that $\mathbf{v}_a \approx \mathbf{v}_b$. Ignoring the last term on the right of the expression (2.276) for the acceleration, we obtain

$$\frac{d\mathbf{v}_a}{dt} \approx \frac{q_a q_b}{4\pi\epsilon_0 m_a \gamma^3} \frac{\mathbf{r}_a - \mathbf{r}_b}{|\mathbf{r}_a - \mathbf{r}_b|^3}. \tag{2.278}$$

This closely resembles the acceleration due to a pure Coulomb force, but is reduced by a factor of $1/\gamma^3$. This reflects the relativistic mass γm_a, and a factor of $1/\gamma^2$ expressing the canceling nature of the electrostatic and magnetic forces at relativistic beam energies.

As the time step Δt is decreased, the particle displacement approaches a stable value. The number of computation steps is inversely proportional to Δt, so one naturally chooses the largest Δt for which the displacement adequately approximates the stable end value. Including more terms in the Taylor expansion improves the convergence in a nonlinear way [88]. It can be shown that the truncation error for a given Δt is inversely proportional to n^m, where n is the number of integration steps, and m is the order of the highest order term in the Taylor series.

One increases the particle sample size N until a stable limiting value of the displacement is obtained. The number of computations for each time step is $N(N-1)/2$. Since this is quadratic in N, it has a large impact on the overall computation time for large N. The length of the simulated beam segment is proportional to N. The force on particles near the ends of the sample is improperly represented. To assess the importance of this, we imagine a half-space filled with charge of some uniform average density. A test particle on the boundary plane between the regions with and without charge experiences a repulsive force due to the charge. Considering the influence of a hemispherical shell of charge on the test particle, the strength of the force is inversely proportional to the square of the radius, and proportional to the total charge in the shell. This latter is proportional to the square of the radius. The net force is thus independent of the radius of the shell, and all shells have the same influence on the test particle, regardless of their radii. The range of the average Coulomb interaction is therefore effectively infinite. It follows from these considerations that the sample length must be much greater than the diameter of the beam for accurate simulation. This can lead to a very large number of particles, and correspondingly long computation time. The stochastic contribution to the displacement is expected to have relatively short range, however, since the fluctuations average out over long distances. A technique exists to take advantage of this by separating the effects of the average and stochastic contributions [40]. The reader is referred to this reference for details.

The above procedure applies to a drift length, with no external fields present. In principle, one can add these fields into the expression for the Lorentz force. For many applications, a thin lens approximation suffices, in which the direction of the velocity is shifted toward the optic axis by an amount proportional to the radial distance off axis. The magnitude of the velocity is unchanged. In this way, complex systems can be analyzed by Monte Carlo simulation of a series of drift lengths separated by thin lenses. Such simulations have shown close agreement with experiment [51, 70]. Monte Carlo simulation thus offers a powerful predictive tool in the design of practical systems.

Problem

Calculate an analytic expression for the time rate of change $\dot{\mathbf{a}}_i$ in terms of the position $\mathbf{r}_i$, velocity $\mathbf{v}_i$, and acceleration $\mathbf{a}_i$.

2.6.2 Analytical approximation by Markov's method of random flights

Monte Carlo simulation is inherently accurate, due to the minimal assumptions needed to describe the physical process. However, it

suffers from two disadvantages. First, it is computationally intensive. For a given system configuration, a new simulation must be
performed for every distinct operating point, and this can be very
time-consuming. Second, it is difficult to achieve an intuitive understanding of the physical process, such as the understanding one
might obtain from an analytical theory.

Given that the N-body problem cannot be solved analytically in
closed form, it is of great interest to inquire whether a suitable analytical approximation can be found. There is a history of attempts
to express beam broadening due to stochastic Coulomb scattering
by analytic approximations. Typically these approximations yield
a simple algebraic dependence on experimental parameters such as
beam energy, system length, beam current, and numerical aperture. The reader is referred to two excellent reviews by Kruit and
Jansen [55] and by Jansen [49] for details.

This approach has the advantage that the optical properties of
a system can be estimated quickly and simply with some degree of
accuracy. This facilitates an intuitive understanding of the dependency on experimental parameters. It has the disadvantage that
the formulas depend on the specific system configuration and on
the operating point.

No simple, general formulation appears to exist. Also, it becomes
necessary to independently evaluate the accuracy of the formula
before relying on it for a detailed design of a system. Assuming the
system has yet to be built, this evaluaton relies on Monte Carlo
simulation. In practice, one uses a judicious combination of analytic approximation and Monte Carlo simulation.

In this study, we attempt an analytical analysis which does not
result in such simple formulas, but strikes at the basic underlying statistical mechanics. The remainder of this section closely
follows the earlier analysis by Groves [41]. We aim to derive a formalism which lends itself well to numerical analysis, where this
analysis is less computationally intensive than Monte Carlo sim-

ulation. To this end, we first discuss of the problem of random flights, originally formulated and solved by Markov, and reviewed by Chandrasekhar [17]. We imagine a general physical process consisting of a number of independent steps of varying size. For example, the process might be the motion of a gas molecule, where the molecule is multiply scattered. Between scattering events, the molecule travels a random distance, which represents the length of a step. We wish to determine the probability that the molecule travels a given net distance after N scattering events. There are many other examples of this general process. A key assumption is that the probability of a single event is independent of past history. Such a succession of events is known as a *Markov chain*.

In mathematical terms, the problem can be stated as follows. We assume the size of the jth step is governed by a probability density $\tau_j(\mathbf{x}_j)$ that the step length will be $\mathbf{x}_j$. Given this, we wish to find the probability density $W_N(\mathbf{X})$ for net displacement $\mathbf{X}$ after N steps with displacements $\mathbf{x}_j$, where $j = 1, \ldots, N$. This is completely general, in that the vector quantities $\mathbf{x}$ and $\mathbf{X}$ can have any dimensionality. The probability $W_N(\mathbf{X})$ is found by integrating over all possible step lengths $\mathbf{x}_j$, subject to the constraint that the individual steps must add up to give the desired displacement $\mathbf{X}$. This is

$$
W_N(\mathbf{X}) = \int \cdots \int \delta \left(\sum_{j=1}^{N} \mathbf{x}_j - \mathbf{X} \right) \tau_1(\mathbf{x}_1) \cdots \tau_N(\mathbf{x}_N) \cdot d\mathbf{x}_1 \cdots d\mathbf{x}_N,
$$

(2.279)

where δ is the Dirac delta function, which ensures that only that space is included in the integration, for which the constraint is met. The delta function has an integral representation given by

$$
\delta \left(\sum_{j=1}^{N} \mathbf{x}_j - \mathbf{X} \right) = \frac{1}{(2\pi)^n} \int d^n\mathbf{k} \exp \left[-i\mathbf{k} \cdot \left(\sum_{j=1}^{N} \mathbf{x}_j - \mathbf{X} \right) \right],
$$

(2.280)

where the integral is performed over the entire n-dimensional space of the n-vector $\mathbf{k}$.

Substituting (2.280) into (2.279), and interchanging order of integrations, we find

$$W_N(\mathbf{X}) = \frac{1}{(2\pi)^n} \int d^n\mathbf{k} \exp(i\mathbf{k} \cdot \mathbf{X})$$

$$\cdot \prod_{j=1}^{N} \left[\int d^n\mathbf{x}_j \exp\left(-i\mathbf{k} \cdot \mathbf{x}_j\right) \tau_j(\mathbf{x}_j) \right]. \quad (2.281)$$

We identify the square bracket as the Fourier transform of $\tau_j(\mathbf{x}_j)$ defined by

$$\tilde{\tau}_j(\mathbf{k}) = \int d^n\mathbf{x}_j \exp\left(-i\mathbf{k} \cdot \mathbf{x}_j\right) \tau_j(\mathbf{x}_j). \quad (2.282)$$

At this point we assume that the same probability $\tau(\mathbf{x})$ governs all individual steps $\mathbf{x}_j$. It is, therefore, permissible to drop the subscript j, yielding

$$W_N(\mathbf{X}) = \frac{1}{(2\pi)^n} \int d^n\mathbf{k} \exp(i\mathbf{k} \cdot \mathbf{X}) \left[\tilde{\tau}(\mathbf{k})\right]^N. \quad (2.283)$$

We identify this as an inverse Fourier transform. This can be abbreviated using a shorthand expression

$$\tilde{W}_N(\mathbf{k}) = \left[\tilde{\tau}(\mathbf{k})\right]^N. \quad (2.284)$$

This represents the general solution to the problem of random flights. It can be understood by applying the convolution theorem of Fourier transforms. This says that the transform of a convolution of two functions is equal to the product of the individual transforms of the functions. In this case, the overall probability is an N-fold convolution of the single-step distribution function with itself, as one would naturally expect. It is evident that the problem is quite naturally expressed in terms of Fourier transforms.

We now seek to apply this general mathematical approach to the specific problem of stochastic Coulomb scattering. We imagine a single particle, chosen at random, intersecting the target plane at some transverse position. This position is determined by the action of the optical system, together with the Coulomb scattering with every other particle. If one could remove the effect of the Coulomb scattering, the particle of interest would intersect the target plane at a different position. We call the vector difference between these two positions the trajectory displacement. It is governed by a probability distribution. In mathematical terms, we wish to find this probability distribution function.

Because the N-body problem cannot be solved in closed form, we are led to seek a suitable approximation. To this end, we imagine a second particle, also chosen at random. The second particle scatters with the first particle, producing a smaller random trajectory displacement. Now we imagine a third particle, chosen at random, producing a small random trajectory displacement of the first particle. Similarly, each of the $N - 1$ particles produces a random displacement of the first particle. Each of these scattering events is a two-body interaction. As such, each event can be solved analytically in principle. We now form the vector sum of all of the $N - 1$ trajectory displacements of the first particle, making a resultant trajectory displacement. This is shown schematically in Figure 2.16. This is essentially the same approximation used as a starting point by Van Leeuwen and Jansen [89], although the details of their analysis are quite different from what is presented here. The vector sum of the two-body displacements is given by $\mathbf{X}_S$, while the N-body displacement is denoted by $\mathbf{X}_N$. In general, these two displacements differ, as they were arrived at by different means.

At this point we form a key hypothesis, namely, the sum of the two-body displacements approximates the N-body displacement to within an error which is small, compared with the displacement. Mathematically, this is expressed as $|\mathbf{X}_N - \mathbf{X}_S|/|\mathbf{X}_N| \ll 1$. This hypothesis can be tested using Monte Carlo simulation. Two separate Monte Carlo simulations are required, one using the N-body algorithm described in the previous section, and another using the vector superposition of two-body interactions described here [4]. The two simulations are run with identical initial conditions for

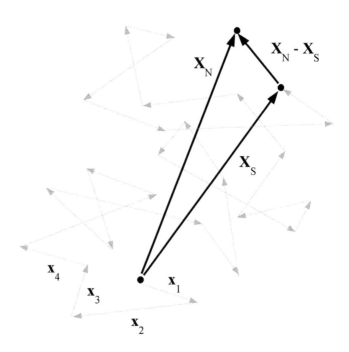

Figure 2.16: N-body interaction and superposition of two-body interactions.

each particle in the sample. The resulting trajectory displacement is compared for each particle individually, with the difference between the two simulations recorded. Statistics are then performed on the differences. It was shown in reference [41] that the two methods agree within about one percent for a particular severe case. The reader is referred to this reference for details. Our hypothesis can safely be considered to be vindicated for a wide range of operating conditions.

The vector superposition of two-body interactions thus represents a useful basis to proceed. This superposition can be formally represented as a Markov chain. This is intuitively evident from Figure 2.16, in which the vector displacements visually appear like a succession of random flights. In fact, the mathematical assumptions are the same, and the above analysis applies. We propose to use the most general possible approach, namely, calculating the trajectory displacement in six-dimensional phase space. The following analysis closely follows [41], with a few minor changes in notation. We define the following quantities:

$$\chi_0 = \text{initial coordinate of an individual particle at time zero}$$
$$\mathbf{s}_0 = \text{initial separation of two particles at time zero}$$
$$\mathbf{s} = \text{separation of two particles at time t, interaction present}$$
$$\mathbf{s}' = \text{separation of two particles at time t, interaction absent,}$$
(2.285)

where all quantities are six-vectors in phase space. The first three components are position, and last three components are momentum. All quantities are random variables, described by probability density functions.

We define an individual particle trajectory displacement as the difference of two-particle separations $\mathbf{s}$ with and $\mathbf{s}'$ without interaction as follows:

$$\epsilon = \tfrac{1}{2}\left(\mathbf{s} - \mathbf{s}'\right),$$
(2.286)

where the factor of $\tfrac{1}{2}$ appears, because the individual particle displacement is half the change in particle-particle separation for particles of equal mass. Consistent with the above hypothesis, we apply the Markov formalism, identifying ϵ with a single Markovian step (2.282). The required Fourier transform for the distribution of trajectory displacements ϵ is

$$\tilde{\tau}(\mathbf{k}) = \int d^6\epsilon \exp(-i\mathbf{k}\cdot\epsilon)\,\tau(\epsilon).$$
(2.287)

We define a probability density $\sigma_0(\chi_0)$ of an initial single-particle six-coordinate χ_0. For example, the beam might be of uniform spatial density, and monoenergetic. In this case the initial distribution $\sigma_0(\chi_0)$ is a constant spatially, multiplied by a delta function in momentum, within the beam volume, and zero outside the beam volume.

Given this, the probability density P_0 of an initial two-particle separation $\mathbf{s}_0$ is

$$P_0(\mathbf{s}_0) = \int d^6\chi_0 \, \sigma_0(\chi_0) \, \sigma_0(\mathbf{s}_0 - \chi_0), \qquad (2.288)$$

where the integral is performed for χ_0 over the initial phase space volume occupied by the beam. The integrand represents the joint probability of finding one particle initially at χ_0, and the second particle displaced by $\mathbf{s}_0$ relative to the first particle. Integrating over all χ_0 ensures that P_0 represents all possible pairs with initial separation $\mathbf{s}_0$.

Alternatively, one could define

$$P_0(\chi_0, \mathbf{s}_0) = \sigma_0(\chi_0) \, \sigma_0(\mathbf{s}_0 - \chi_0). \qquad (2.289)$$

This would retain the correlation with absolute single-particle six-coordinate χ_0. The first case, in which we integrate over χ_0 leads by definiton to just the stochastic interaction arising from local fluctuations in the charge density. The second case without integration leads to the full result. This includes both the stochastic interaction and the systematic effects arising from the global charge distribution within the beam. For brevity in the following, we consider the first case only.

We now make use of the fact that trajectories are conserved in phase space. In any small volume of phase space, this is expressed as

$$d^6N = P_0(\mathbf{s}_0) \, d^6\mathbf{s}_0 = P(\mathbf{s}) \, d^6\mathbf{s} = P'(\mathbf{s}') \, d^6\mathbf{s}' = \tau(\epsilon) \, d^6\epsilon. \qquad (2.290)$$

It follows that (2.287, 2.290):

$$\tilde{\tau}(\mathbf{k}) = \int d^6\mathbf{s}_0 \, P_0(\mathbf{s}_0) \exp(-i\mathbf{k} \cdot \epsilon), \qquad (2.291)$$

where the integral is performed over the space of initial two-particle separations, determined by (2.288). The two-body scattering has an analytic solution for the separation $\mathbf{s}$ in terms of the initial condition $\mathbf{s}_0$. Separately, the separation $\mathbf{s}'$ in the absence of interaction is also determined analytically from $\mathbf{s}_0$. From (2.286) the trajectory displacement ϵ is thus found in terms of $\mathbf{s}_0$. This permits us to perform the integral (2.291). Substituting this into (2.283), we obtain the general solution for the stochastic Coulomb interaction, in the vector superposition of two-body interactions.

We are often interested in only the transverse coordinates in some target plane, or alternatively, the broadening of kinetic energy, for example. In these cases, the other degrees of freedom, such as the displacement in the axial coordinate, are superfluous. We need to integrate over all of the superfluous degrees of freedom. Fortunately, the form of (2.287) makes this particularly simple. Due to a theorem of Fourier transforms, setting the frequency k equal to zero is equivalent to integrating over the entire range of the variable in direct space. By setting the superfluous components of $\mathbf{k}$ equal to zero, we automatically integrate over these degrees of freedom in the direct space of ϵ. This leaves only those degrees of freedom we are interested in. In particular, this applies to (2.291), where the superfluous degrees of freedom integrate to unity.

The general solution for the trajectory displacement is given by (2.283, 2.291). In general, the integral in (2.291) must be performed numerically. A special case exists for which a closed-form analytic solution exists, however. This is the case in which all particles are initially at rest in the rest frame of the beam particles. This is equivalent to an initially monoenergetic beam with zero energy in the rest frame. It follows that the beam is monoenergetic in the lab frame as well. It is simpler to perform the calculation in the rest frame, as the magnetic Lorentz force is zero, and the choice of reference frame does not affect the transverse position. The initial condition for the particle separation in six dimensions is

$$P_0(\mathbf{s}_0) = \psi(\mathbf{r}_0) \cdot \delta(\mathbf{p}_0), \qquad (2.292)$$

where $\mathbf{r}_0$ is the initial particle spatial separation, and $\mathbf{p}_0$ is initial momentum difference. The spatial separation distribution $\psi(\mathbf{r}_0)$ will be determined later. Integrating over all momenta by setting the momentum components of $\mathbf{k}$ to zero, we find (2.291)

$$\tilde{\tau}(\mathbf{k}_r; 0) = \int_0^\infty d\rho_0\, \rho_0 \int_{-\infty}^\infty dz_0\, \psi(\rho_0, z_0) \int_0^{2\pi} d\phi_0 \exp\left(-i\mathbf{k}_r \cdot \epsilon_r\right),$$

(2.293)

where $\mathbf{k}_r$ is the three-vector spatial part of the six-vector $\mathbf{k}$. We have set the three-vector momentum part $\mathbf{k}_p$ to zero. This is equivalent to integrating over all momentum values in direct space.

Next we must find the spatial trajectory displacement ϵ_r at time t. For the case in which the two particles are initially at rest, the scattering reduces to the Kepler problem for zero angular momentum, where the particles fly apart along a line joining them. In this case the solution to the Kepler problem reduces to

$$t\sqrt{\frac{e^2}{\pi\epsilon_0 m r_0^3}} = \frac{r}{r_0}\sqrt{1 - \frac{r_0}{r}} + \tanh^{-1}\sqrt{1 - \frac{r_0}{r}},$$

(2.294)

where t is transit time, r_0 is initial separation, and r is separation at time t. Substituting this into (2.293) we obtain

$$\tilde{\tau}(\mathbf{k}_\rho, 0; 0) = 2\pi \int_0^\infty d\rho_0\, \rho_0 \int_{-\infty}^\infty dz_0\, \psi(\rho_0, z_0)\, J_0\left[\tfrac{1}{2} k_\rho \rho_0 \left(\frac{r}{r_0} - 1\right)\right],$$

(2.295)

where $\mathbf{k}_\rho$ is the two-vector spatial part in the transverse (ρ, ϕ) plane. We have made use of assumed axial symmetry and the integral representation of the Bessel function $J_0(x)$ as

$$J_0(x) = \frac{1}{2\pi} \int_0^{2\pi} d\phi\, e^{-ix\cos\phi}.$$

(2.296)

We assume the particles are initially uniformly distributed over a cylindrical volume with radius a and length L. It follows that $\psi_0(\mathbf{r}_0)$ is convolution of a cylindrical volume with itself, as follows:

$$\psi_0(\mathbf{r}_0) = \frac{1}{\pi a^2 L} \cdot \frac{2}{\pi}\left[\cos^{-1}\left(\frac{\rho_0}{2a}\right) - \frac{\rho_0}{2a}\sqrt{1 - \left(\frac{\rho_0}{2a}\right)^2}\right] \cdot \left(1 - \frac{|z_0|}{L}\right),$$

(2.297)

where ψ_0 is nonzero for

$$0 \le \rho_0 \le 2a, \quad -L \le z_0 \le L,$$

(2.298)

and zero outside this region. This yields the probability density for $N - 1$ scattering particles:

$$W_N(\epsilon_{N\rho}) = \frac{1}{2\pi} \int_0^\infty dk_\rho \, k_\rho \, [\tilde{\tau}\,(\mathbf{k}_\rho, 0; 0)]^{N-1} J_0\,(k_\rho\,\epsilon_{N\rho}), \quad (2.299)$$

where $\epsilon_{N\rho}$ is magnitude of transverse component of resultant (net) trajectory displacement, and we have made use of axial symmetry. Equations (2.295, 2.297, 2.299) represent the solution. It is shown in [41] that this solution agrees quantitatively with Monte Carlo simulation for a particular severe case.

The main result of this section is contained in equations (2.283, 2.291) for the six-vector trajectory displacement in phase space. In general, the integral in (2.291) must be performed numerically, given an assumed form for the initial distribution coordinates $\sigma_0(\chi_0)$. We have further shown that the dimensionality can be reduced in a straightforward manner by framing the problem in terms of Fourier transforms. This has enormous practical significance for extracting quantitative values for the components of trajectory displacement.

2.7 Hamilton–Jacobi theory

The solution to the general dynamical problem in classical mechanics follows directly from Hamilton's principle of least action (2.8). An alternative, but completely equivalent formulation exists, in the theory due to Hamilton and Jacobi. This formulation will prove useful in the following chapter, where we will explore the correspondence between the classical and quantum mechanical descriptions of single-particle motion in the presence of a general electromagnetic potential. This section closely follows the analysis of Goldstein, et. al. [35].

2.7.1   Canonical transformations

We begin with Hamilton's equations of motion (2.85), which are given in their most general form as

$$\frac{\partial H}{\partial P_i} = \dot{Q}_i, \qquad \frac{\partial H}{\partial Q_i} = -\dot{P}_i, \qquad i = 1, \ldots, n, \qquad (2.300)$$

where the Q_i are the n generalized coordinates, and P_i are the n conjugate canonical momenta. The Hamiltonian is a function of all coordinates and momenta, where

$$H = H(Q_1, \ldots, Q_n; P_1, \ldots, P_n; t). \qquad (2.301)$$

In general, H can have explicit dependence on the time t, which we regard as a parameter which uniquely specifies a given point along the trajectory. For brevity, we adopt a vector notation where $H = H(\mathbf{Q}, \mathbf{P}; t)$, and

$$\begin{aligned} \mathbf{Q} &= (Q_1, \ldots, Q_n) \\ \mathbf{P} &= (P_1, \ldots, P_n). \end{aligned} \qquad (2.302)$$

In principle, Hamilton's equations can be integrated to find the coordinates $Q_i(t)$ and canonical momenta $P_i(t)$ as functions of the time t. This would constitute a formal solution to the general dynamical problem.

It is always possible to transform to a new system of coordinates and momenta. As a simple example, one could transform from Cartesian to spherical coordinates. This would simplify a problem with spherical symmetry, such as scattering by a spherically symmetric Coulomb potential. The components of canonical momentum would also transform to a spherical system.

In this context, we imagine a transformation to a system where all coordinates and momenta are constants of the motion. If such a transformation were possible, this would represent an immediate formal solution to the general dynamical problem, since the coordinates and momenta would simply be equal to their initial values at time zero. Assuming for the moment that such a transformation can be found, we express the new set

$$q_i = q_i(\mathbf{Q}, \mathbf{P}), \qquad p_i = p_i(\mathbf{Q}, \mathbf{P}), \qquad i = 1, \ldots, n, \qquad (2.303)$$

where the new coordinates q_i and canonical momenta p_i are as yet unspecified functions of the old Q_i and P_i. Here we adopt a different notation from the earlier sections, for reasons which will become clear in the following. The p_i are not to be confused with the components of kinetic momentum described earlier. In order for the motion to be physically possible, we require that the new q_i, p_i also obey Hamilton's equations

$$\frac{\partial K}{\partial p_i} = \dot{q}_i, \qquad \frac{\partial K}{\partial q_i} = -\dot{p}_i, \qquad i = 1, \ldots, n, \qquad (2.304)$$

where $K(\mathbf{q}, \mathbf{p}; t)$ is the Hamiltonian in the new system. Any transformation for which Hamilton's equations of motion (2.300, 2.304) are satisfied in both the old and new systems is called a *canonical transformation*.

Hamilton's principle (2.8) can be written separately in the two systems (2.19) as

$$\delta \int_{t_1}^{t_2} \left[\sum_i P_i \dot{Q}_i - H(\mathbf{Q}, \mathbf{P}; t) \right] dt = 0$$

$$\delta \int_{t_1}^{t_2} \left[\sum_i p_i \dot{q}_i - K(\mathbf{q}, \mathbf{p}; t) \right] dt = 0. \qquad (2.305)$$

In order for both equations to hold, the integrands can differ at most by the total time derivative of an arbitrary function F as follows:

$$\sum_i P_i \dot{Q}_i - H = \sum_i p_i \dot{q}_i - K + \frac{d}{dt} F(\mathbf{Q}, \mathbf{q}; t), \qquad (2.306)$$

where this represents a necessary condition relating the Hamiltonians H and K in the two systems. The function F is called a *generating function.* Expanding dF/dt by the chain rule for partial derivatives, we find

$$\frac{dF}{dt} = \sum_{i=1}^{n} \left[\frac{\partial F}{\partial Q_i} \dot{Q}_i + \frac{\partial F}{\partial q_i} \dot{q}_i \right] + \frac{\partial F}{\partial t}. \qquad (2.307)$$

Equating coefficients of $\dot{Q}$ and $\dot{q}$ respectively, we obtain

$$P_i = \frac{\partial F}{\partial Q_i}, \qquad p_i = -\frac{\partial F}{\partial q_i}, \qquad K = H + \frac{\partial F}{\partial t}. \qquad (2.308)$$

Other generating functions can be constructed. For example, we can define a new function S by

$$S(\mathbf{Q}, \mathbf{p}; t) = F(\mathbf{Q}, \mathbf{q}; t) + \sum_i p_i q_i. \qquad (2.309)$$

This is an example of a Legendre transformation. Substituting, it follows that

$$P_i = \frac{\partial S}{\partial Q_i}, \qquad q_i = \frac{\partial S}{\partial p_i}, \qquad K = H + \frac{\partial S}{\partial t}. \qquad (2.310)$$

At this point we make a key assumption: we imagine a transformation for which $K \equiv 0$, i.e., the Hamiltonian K in the new system is identically zero. Assuming such a transformation can be found, it would follow from Hamilton's equations of motion in the new system that

$$\frac{\partial K}{\partial p_i} = \dot{q}_i = 0, \qquad \frac{\partial K}{\partial q_i} = -\dot{p}_i = 0, \qquad i = 1, \ldots, n. \qquad (2.311)$$

From this we would immediately find that

$$p_i = \alpha_i = const, \qquad q_i = \beta_i = const, \qquad i = 1, \ldots, n, \qquad (2.312)$$

consistent with our original intent. The α_i and β_i constitute $2n$ integration constants. Substituting above, we find

$$H\left(Q_1, \ldots, Q_n, \frac{\partial S}{\partial Q_1}, \ldots, \frac{\partial S}{\partial Q_n}; t \right) + \frac{\partial S}{\partial t} = 0. \qquad (2.313)$$

This equation is called the Hamilton-Jacobi equation, and S is called *Hamilton's principal function*. We notice the significant fact that this equation only contains the variables Q_i and t.

This immediately leads to a formal procedure to solve the dynamical problem in principle: we substitute $\partial S/\partial Q_i$ for P_i in H, then integrate to solve for $S(\mathbf{Q}, \alpha; t)$. With S known, we then invert the equation

$$\beta_i = \frac{\partial}{\partial \alpha_i} S(\mathbf{Q}, \alpha; t) \tag{2.314}$$

to solve for the $Q_i(t)$; i.e., find $Q_i = Q_i(\alpha, \beta; t)$. The α_i and β_i are determined by the initial conditions. This represents a formal solution to the general dynamical problem.

It is of particular interest to consider the important special case where the original Hamiltonian H has no explicit time dependence. The above procedure in (2.305) to (2.308) applies, with the difference that the time t does not appear explicitly in $H(\mathbf{Q}, \mathbf{P})$ and $K(\mathbf{q}, \mathbf{p})$. It follows from this that the generating function $F(\mathbf{Q}, \mathbf{q})$ has no explicit time dependence, and

$$\frac{\partial F}{\partial t} = 0. \tag{2.315}$$

We now define a new generating function $W(\mathbf{Q}, \mathbf{p})$ as

$$W(\mathbf{Q}, \mathbf{p}) = F(\mathbf{Q}, \mathbf{q}) + \sum_{i=1}^{n} p_i \, q_i. \tag{2.316}$$

From (2.307, 2.315, 2.316) it follows that

$$P_i = \frac{\partial W}{\partial Q_i}, \qquad q_i = \frac{\partial W}{\partial p_i}, \qquad K = H. \tag{2.317}$$

Hamilton's equations in the transformed system are

$$\frac{\partial K}{\partial p_i} = \dot{q}_i, \qquad \frac{\partial K}{\partial q_i} = -\dot{p}_i, \qquad i = 1, \ldots, n. \tag{2.318}$$

We now make the key assumption that

$$H = K = \alpha_1 = p_1 = const, \qquad (2.319)$$

where H is the conserved total energy, which we thus identify with the first conserved component of the new canonical momentum p_1. It follows immediately that

$$\frac{\partial K}{\partial q_i} = -\dot{p}_i = 0, \qquad p_i = \alpha_i = const, \qquad (2.320)$$

where the α_i form n integration constants. Also, from (2.318, 2.319),

$$\frac{\partial K}{\partial p_i} = \dot{q}_i = \begin{cases} 1 & \text{when } i = 1, \\ 0 & \text{when } i = 2, \ldots, n. \end{cases} \qquad (2.321)$$

This leads to

$$q_i = \begin{cases} t + \beta_1 & \text{when } i = 1, \\ \beta_i & \text{when } i = 2, \ldots, n, \end{cases} \qquad (2.322)$$

where the β_i form n integration constants. From (2.319) it follows that

$$H\left(Q_1, \ldots, Q_n, \frac{\partial W}{\partial Q_1}, \ldots, \frac{\partial W}{\partial Q_n}\right) = \alpha_1. \qquad (2.323)$$

This is the Hamilton-Jacobi equation for the special case where the Hamiltonian H has no explicit time dependence. The function W is called *Hamilton's characteristic function*. Also, it follows that

$$\beta_i = \begin{aligned} & \frac{\partial}{\partial \alpha_i} W(\mathbf{Q}, \alpha) - t, & i = 1 \\[2mm] & \frac{\partial}{\partial \alpha_i} W(\mathbf{Q}, \alpha), & i = 2, \ldots, n. \end{aligned} \qquad (2.324)$$

As before, this immediately leads to a formal procedure to solve the dynamical problem in principle: we substitute $\partial W/\partial Q_i$ for P_i in H, then integrate to solve for $W(\mathbf{Q}, \alpha)$. With W known, we then invert the equation

$$\beta_i = \frac{\partial}{\partial \alpha_i} W(\mathbf{Q}, \alpha) \qquad (2.325)$$

to solve for the $Q_i(t)$; i.e., find $Q_i = Q_i(\alpha, \beta)$. The α_i and β_i are determined by the initial conditions. This represents the solution.

The transformations generated by W and S have quite different properties. From the third of equations (2.310) and from equation (2.319), the two generating functions are related by

$$S(\mathbf{Q}, \mathbf{p};\, t) = W(\mathbf{Q}, \mathbf{p}) - \alpha_1 t \tag{2.326}$$

for the case where H has no explicit time dependence. This completes the formal solution to the dynamical problem by Hamilton-Jacobi theory.

It is interesting to explore the relationship between the generating functions S and W, and Hamilton's principle of least action (2.8). We begin by making a working hypothesis, namely, Hamilton's principle function S can be written as an indefinite integral

$$S(\mathbf{Q},\, \alpha;\, t) = \int^{t} L(\mathbf{Q},\, \dot{\mathbf{Q}},\, t')\, dt', \tag{2.327}$$

where L is the Lagrangian. We now proceed to test the validity of this hypothesis. From the definition of the Hamiltonian (2.19) we rewrite this as

$$S(\mathbf{Q},\, \alpha;\, t) = \int^{t} \left[\sum_{i=1}^{n} P_i \dot{Q}_i - H(\mathbf{Q}, \mathbf{P};\, t') \right] dt'. \tag{2.328}$$

Taking the partial derivative with respect to Q_i, we find

$$\frac{\partial S}{\partial Q_i} = \int^{t} dt'\, \frac{\partial}{\partial Q_i} \left(\sum_{i=1}^{n} P_i \dot{Q}_i \right) - \int^{t} dt'\, \frac{\partial H}{\partial Q_i}. \tag{2.329}$$

The first term on the right is identically zero, since the quantity in large parentheses has no dependence on Q_i. From the second of Hamilton's equations (2.300), the second term on the right is equal to P_i, giving

$$\frac{\partial S}{\partial Q_i} = P_i, \tag{2.330}$$

identical with the first equation (2.310). Taking the partial derivative with respect to time, we obtain

$$\frac{\partial S}{\partial t} = \frac{\partial}{\partial t} \int \left(\sum_{i=1}^{n} P_i \, dQ_i \right) - \frac{\partial}{\partial t} \int^t H(\mathbf{Q}, \mathbf{P}; t') \, dt'. \qquad (2.331)$$

The first term on the right is identically zero, since the integral has no explicit time dependence. The second term on the right is $H(\mathbf{Q}, \mathbf{P}; t)$. This leads to

$$H + \frac{\partial S}{\partial t} = 0, \qquad (2.332)$$

identical with (2.313). We assumed in (2.327) that S is an indefinite integral, and is therefore defined only to within an additive constant. This integration constant can always be chosen in principle so that (2.314) is satisfied, remembering that the α_i and β_i are constants of the motion. This completes the justification of our postulate (2.327) for the form of S. We have thus identified Hamilton's principal function S with the indefinite integral corresponding to the action integral in Hamilton's principle of least action (2.8).

Next we consider the case where the Hamiltonian $H(\mathbf{Q}, \mathbf{P})$ has no explicit time dependence. We rewrite (2.328) as

$$S(\mathbf{Q}, \alpha; t) = \int \mathbf{p} \cdot d\mathbf{Q} - H t, \qquad (2.333)$$

remembering that $H = \alpha_1$ is the constant total energy. From (2.326), it follows that

$$W(\mathbf{Q}, \alpha) = \int \mathbf{P} \cdot d\mathbf{s}, \qquad (2.334)$$

where the right side is the indefinite path integral along the physical trajectory. We have thus identified Hamilton's characteristic function W with the indefinite integral corresponding to the action integral in the mechanical equivalent of Fermat's principle (2.41).

2.7.2 Applications of Hamilton–Jacobi theory

To convey a feeling for how to use of the theory, and to show that it actually works, we now apply it to two well-known examples [35].

The first example is the one-dimensional harmonic oscillator. The Hamiltonian is

$$H = \frac{P^2}{2m} + \frac{kQ^2}{2}, \tag{2.335}$$

where k is the spring constant, and $Q =$ is the coordinate for the displacement. Obviously, the Hamiltonian H has no explicit time dependence. According to our prescription, we now substitute for the momentum $P = \partial W / \partial Q$. The resulting Hamilton-Jacobi equation is

$$\frac{1}{2m} \left(\frac{\partial W}{\partial Q} \right)^2 + \frac{kQ^2}{2} = \alpha_1, \tag{2.336}$$

where α_1 is the conserved total energy. Hamilton's characteristic function W is expressed as the integral

$$W(Q, \alpha_1) = \int dQ \sqrt{2m\alpha_1 - mkQ^2}. \tag{2.337}$$

Also,

$$\beta_1 = \frac{\partial W}{\partial \alpha_1} - t. \tag{2.338}$$

Substituting, we obtain

$$t + \beta_1 = \sqrt{\frac{m}{2\alpha_1}} \int \frac{dQ}{\sqrt{1 - k\,Q^2/(2\alpha_1)}} = -\sqrt{\frac{m}{k}} \arccos\left(\sqrt{\frac{k}{2\alpha_1}} Q \right). \tag{2.339}$$

We invert this to solve for Q as follows:

$$Q(\alpha_1, \beta_1) = \sqrt{\frac{2\alpha_1}{k}} \cos\left[\sqrt{\frac{k}{m}} (t + \beta_1) \right]. \tag{2.340}$$

We now assume an initial condition that the displacement Q is at its maximum Q_0 at $t = 0$. From this it fillows

$$\beta_1 = 0, \qquad \alpha_1 = \tfrac{1}{2} k\, Q_0^2, \tag{2.341}$$

where α_1 is the total conserved energy. This is equal to the potential energy at maximum displacement Q_0, where the kinetic energy is zero. Also,

$$Q = Q_0 \cos\left(\sqrt{\frac{k}{m}}\, t\right).\qquad(2.342)$$

This represents the solution, expressing the familiar cosinusoidal motion, where $\sqrt{k/m}$ is the angular frequency.

As a second example, we study the Kepler problem. This will have additional significance in classical Rutherford scattering, which will be explored in Chapter 4. The Hamiltonian is

$$H = \frac{1}{2m}\left(P_r^2 + \frac{P_\theta^2}{r^2}\right) + U(r),\qquad(2.343)$$

where U is the potential energy. Hamilton's equation for the angular momentum is

$$\frac{\partial H}{\partial \theta} = -\dot{P}_\theta = 0,\qquad P_\theta = \alpha_2 = const.\qquad(2.344)$$

We write the radial and angular momenta as

$$P_r = \frac{\partial W}{\partial r},\qquad P_\theta = \frac{\partial W}{\partial \theta} = \alpha_2.\qquad(2.345)$$

This leads to a separable form for W as follows:

$$W(r,\theta,\alpha_1,\alpha_2) = W_r(r,\alpha_1,\alpha_2) + \alpha_2\theta.\qquad(2.346)$$

The Hamilton-Jacobi equation is

$$\frac{1}{2m}\left[\left(\frac{\partial W}{\partial r}\right)^2 + \frac{\alpha_2^2}{r^2}\right] + U(r) = \alpha_1.\qquad(2.347)$$

Rearranging terms and taking the square root of both sides, we find

$$\frac{\partial W_r}{\partial r} = \sqrt{2m(\alpha_1 - U) - \frac{\alpha_2^2}{r^2}}.\qquad(2.348)$$

Hamilton's characteristic function W is thus expressed as the integral

$$W = \int dr \sqrt{2m(\alpha_1 - U) - \frac{\alpha_2^2}{r^2}} + \alpha_2\theta. \tag{2.349}$$

Also,

$$\beta_1 = \frac{\partial W}{\partial \alpha_1} - t. \tag{2.350}$$

This is equivalent to

$$t + \beta_1 = m \int \frac{dr}{\sqrt{2m(\alpha_1 - U) - \alpha_2^2/r^2}}. \tag{2.351}$$

Furthermore,

$$\beta_2 = \frac{\partial W}{\partial \alpha_2}, \tag{2.352}$$

and

$$\theta - \beta_2 = -\alpha_2 \int \frac{dr}{r^2\sqrt{2m(\alpha_1 - U) - \alpha_2^2/r^2}}. \tag{2.353}$$

To this point we have not yet specified a precise form for the radially symmetric potential $U(r)$, and the analysis remains general in this regard.

At this point we assume an inverse law for U, namely

$$U(r) = \frac{\kappa}{r}, \tag{2.354}$$

where κ is a real constant. The Coulomb force between two charges q_1 and q_2 has

$$\kappa = \frac{q_1 q_2}{4\pi\epsilon_0}, \tag{2.355}$$

for example. For charges of like sign, $\kappa > 0$, and the force is repulsive. For charges of opposite sign, $\kappa < 0$, and the force is attractive. Making a change of variables $u = 1/r$, the equation for θ is immediately integrated to give

$$\theta - \beta_2 = -\cos^{-1}\left[\frac{\alpha_2^2 u + m\kappa}{\sqrt{m^2\kappa^2 + 2m\alpha_1\alpha_2^2}} \right]. \tag{2.356}$$

We invert this to solve for $u = 1/r$ as a function of θ as follows:

$$\frac{1}{r} = -\frac{m\kappa}{P_\theta^2}\left[1 + \sqrt{1 + \frac{2HP_\theta^2}{m\kappa^2}}\cos(\theta - \theta_0)\right]. \qquad (2.357)$$

We have identified

$$\alpha_1 = H, \qquad \alpha_2 = P_\theta, \qquad \beta_2 = \theta_0, \qquad (2.358)$$

where H is the conserved total energy, and P_θ is the conserved angular momentum. We define a quantity called the *eccentricity* as

$$\epsilon = \sqrt{1 + \frac{2HP_\theta^2}{m\kappa^2}}. \qquad (2.359)$$

For $0 < \epsilon < 1$ the orbit is an ellipse, for $\epsilon = 1$ it is a parabola, and for $\epsilon > 1$ it is a hyperbola. The integral (2.351) for t cannot be expressed in closed form, but we assume an initial condition $\beta_1 = -t_0$.

The main result is the orbit equation (2.357). This will apply directly to classical Rutherford scattering in Chapter 4.

2.7.3 Hamilton–Jacobi theory and geometrical optics

Adopting the notation of earlier sections, the non-relativistic form of the conserved Hamiltonian H is

$$H = \frac{p^2}{2m} + U(\mathbf{x}) = const, \qquad (2.360)$$

where $U(\mathbf{x}) = q\phi(\mathbf{x})$ is the time independent potential energy, and q is the charge of the particle. The canonical momentum $\mathbf{P}$ is given by

$$\mathbf{P} = \mathbf{p} + q\,\mathbf{A}, \qquad (2.361)$$

where **p** is the kinetic momentum and **A** is the magnetic vector potential. According to the preceding analysis, Hamiton's characteristic function W is related to the canonical momentum **P** by

$$P_i = \frac{\partial W}{\partial x_i}. \tag{2.362}$$

The Hamilton-Jacobi equation is

$$\sum_i \left(\frac{\partial W}{\partial x_i} - q\, A_i \right)^2 = 2\, m\, (\alpha_1 - q\, \phi), \tag{2.363}$$

where $\alpha_1 = H$ is the conserved total energy, and the right side is the square of the scalar kinetic momentum. In principle, this can be solved for the trajectory by the above procedure, but no simple, closed-form solution exists.

This is related to the action integral W_{ab} by

$$\int_{\mathbf{x}_a}^{\mathbf{x}_b} \mathbf{P} \cdot d\mathbf{s} = \int_{\mathbf{x}_a}^{\mathbf{x}_b} \nabla W \cdot d\mathbf{s} = \int_{\mathbf{x}_a}^{\mathbf{x}_b} \frac{\partial W}{\partial s}\, ds = \left[W(\mathbf{x}) \right]_{\mathbf{x}_a}^{\mathbf{x}_b} \equiv W_{ab},$$
$$\tag{2.364}$$

where the integration path corresponds to a physical ray if and only if W satisfies the Hamilton-Jacobi equation. The optical path length W_{ab} is identical with Hamilton's characteristic function evaluated between the two end points $\mathbf{x}_a$ and $\mathbf{x}_b$. We have made use of

$$\mathbf{P} = \nabla W, \tag{2.365}$$

which means that the canonical momentum **P** is normal to the surfaces $W = const$ along the ray path. In the case where $\mathbf{A} = 0$ (no magnetic field), the kinetic momentum **p** is normal to the surfaces $W = const$.

A relativistic generalization for a hypothetical spin-zero particle can be formed from

$$H = \sqrt{p^2 c^2 + m^2 c^4} + q\, \phi = const, \tag{2.366}$$

which leads to the Hamilton–Jacobi equation

$$\sqrt{(\nabla W - q\, \mathbf{A})^2 c^2 + m^2 c^4} + q\phi = \alpha_1. \tag{2.367}$$

The equations (2.363) and (2.367) represent the main results of this section. They are not amenable to closed-form analytical solution, but will have far-reaching consequences in the correspondence between the classical and quantum mechanical analyses to come in Chapter 3.

Chapter 3

Wave optics

To this point we have discussed the geometrical optics of charged particles in electric and magnetic fields, based on relativistic classical mechanics. This fails to explain the important class of phenomena arising from diffraction and interference of matter waves. A proper description begins with quantum mechanics.

This is perhaps best appreciated by considering the analogy with light optics. Einstein's original hypothesis in 1905 holds that the electromagnetic field is quantized. As such, light propagates in discrete energy packets called *photons*. Furthermore, acording to a later hypothesis by Einstein, a single photon is endowed with momentum p which satisfies

$$p = \frac{h}{\lambda}, \tag{3.1}$$

where h is Planck's constant, given by $h = 6.6261 \times 10^{-34}$ Joule-sec, and λ is the wavelength.

A later hypothesis of de Broglie states that this same relationship between momentum and wavelength holds for a particle with mass and charge. This indicates a close analogy between the dynamical motion of a charged particle and a photon. Both exhibit particle- and wave-like behavior.

Interference involves a single particle or photon propagating over alternative paths, with an associated uncertainty in the path taken. Position and momentum are described by a complex amplitude or wave function, with the amplitudes for alternative paths adding to form a resultant complex amplitude. The absolute square of this amplitude gives the probability or intensity. We will explore the wave function in detail in the following sections. The spatial part of the wave equation governing the propagation of this amplitude is the same for a free particle and a free photon. Consequently, the formalism of scalar diffraction, interference, and image formation for light can be directly applied to particles.

The path taken by a ray of light can be found from Fermat's principle, which states that the physical path represents the shortest possible transit time through a medium. The path taken by a particle can be found from the principle of least action, which states that the physical path represents the minimum of the action integral. These two principles are strikingly similar. They arose from a classical description, but as we shall see in the following, each has a wave-optical analog as well. No classical analog exists for the quantum mechanical description. However, quantum mechanical motion of particles and photons approaches classical behavior in the high-energy limit. The analogy between particle optics and light optics is deep and pervasive.

The central problem in this chapter is to solve for the wave function for a particle of charge q and rest mass m propagating in a general electromagnetic potential. With this foundation, we then explore a few of the important implications for wave optics. We begin with a review of basic quantum mechanics governing particle motion. We confine the discussion to only those topics which are relevant to the motion of a fast (unbound) charged particle in a general electromagnetic potential. This is the subject of the following section.

3.1 Quantum mechanical description of particle motion

We seek a dynamical equation to describe the motion of a single particle of charge q and rest mass m in the presence of a general electromagnetic potential. To this end, we begin with a review of basic quantum mechanics. For clarity, we will do this deductively, beginning with the fundamental postulates of quantum mechanics, and proceeding to the motion of a single charged particle in a general electromagnetic potential. The reader can refer to any of a number of excellent textbooks on basic quantum mechanics. [59], [79].

3.1.1 The postulates of quantum mechanics

We begin with a fundamental postulate as follows:

Every measurable dynamical variable C has a corresponding operator $\hat{C}$, which satisfies a linear operator equation

$$\hat{C}\varphi = c\varphi. \tag{3.2}$$

The dynamical variable C can be any measurable physical quantity. Examples include position, momentum, and energy, to name a few. The operator $\hat{C}$ acts on the function φ, which is called an *eigenfunction*. The multiplicative constant c is called an *eigenvalue*. In the following we will always denote an operator by a hat over the letter, to distinguish it from an ordinary variable or function. The definiton of a linear operator is explored further in Problem 1.

The eigenfunction and eigenvalue are not necessarily unique, but can take on various values. The number and character of possible values depends on the physical situation being described. For clarity, we therefore rewrite the operator equation (3.2) as

$$\hat{C}\,\varphi_j = c_j\,\varphi_j, \tag{3.3}$$

where the subscript j labels the particular eigenfunction φ_j and its corresponding eigenvalue c_j.

At this point we state a second postulate as follows:

A single precise measurement of the dynamical variable C yields one and only one of the eigenvalues c_j.

This postulate establishes the physical significance of the eigenvalues c_j, namely, each eigenvalue is a possible result of a measurement of the corresponding dynamical variable. The physical significance of the eigenfunctions φ_j will be made clear later.

We now proceed to apply this formalism to the motion of a charged particle. We begin by defining the operators which correspond to the dynamical variables of interest. The operators corresponding to the three Cartesian coordinates of position **x** are defined as

$$\begin{aligned}
\hat{x} &= x \\
\hat{y} &= y \\
\hat{z} &= z,
\end{aligned} \tag{3.4}$$

where the operation is multiplication. In words, the operators corresponding to the Cartesian coordinates are the coordinates themselves.

The operators corresponding to the three Cartesian components of the magnetic vector potential **A**(**x**), and to the electrostatic potential $\phi(\mathbf{x})$ are defined, respectively, as

$$\begin{aligned}
\hat{A}_x &= A_x(\mathbf{x}) \\
\hat{A}_y &= A_y(\mathbf{x}) \\
\hat{A}_z &= A_z(\mathbf{x}) \\
\hat{\phi} &= \phi(\mathbf{x}),
\end{aligned} \tag{3.5}$$

where the operation is again multiplication. We assume for now that the electromagnetic potentials have no explicit time dependence. We will generalize this to the time-dependent case in a later section.

The operators corresponding to the three Cartesian components of the classical canonical momentum are defined as

$$
\begin{aligned}
\hat{P}_x &= -i\hbar \frac{\partial}{\partial x} \\
\hat{P}_y &= -i\hbar \frac{\partial}{\partial y} \\
\hat{P}_z &= -i\hbar \frac{\partial}{\partial z},
\end{aligned}
\tag{3.6}
$$

where the operation is partial differentiation. The operators corresponding to the three Cartesian components of the kinetic momentum are defined as

$$
\begin{aligned}
\hat{p}_x &= \hat{P}_x - q\,\hat{A}_x \\
\hat{p}_x &= \hat{P}_x - q\,\hat{A}_x \\
\hat{p}_x &= \hat{P}_x - q\,\hat{A}_x
\end{aligned}
$$

$$
\tag{3.7}
$$

by analogy with the classical definition (2.25), where q is the charge of the particle.

Finally, the operator corresponding to the classical Hamiltonian function H is defined as

$$
\hat{H} = i\hbar \frac{\partial}{\partial t},
\tag{3.8}
$$

where t is the time, and the operation is partial differentiation.

Although we discuss Cartesian coordinates, this description can be made to apply to different types of coordinate systems. The discussion is quite general in this respect. We will continue to use Cartesian coordinates here, because an intuitive picture emerges which does not depend strictly on the choice of coordinates.

We now seek an operator equation which describes the evolution of the particle motion in quantum mechanical terms. The classical conserved Hamiltonian is given in the nonrelativistic limit by

$$H = \frac{1}{2m} \left(p_x^2 + p_y^2 + p_z^2 \right) + q\phi(\mathbf{x}). \tag{3.9}$$

At this point we make a fundamental assumption, namely,

A valid operator equation can be constructed by substituting the operator expressions for every classical quantity in the dynamical equation.

Refering to (3.9), This gives

$$\hat{H}\,\psi(\mathbf{x}, t) = \left[\frac{1}{2m} \left(\hat{p}_x^2 + \hat{p}_y^2 + \hat{p}_z^2 \right) + q\hat{\phi}(\mathbf{x}) \right] \psi(\mathbf{x}, t). \tag{3.10}$$

Adhering to our description, $\psi(\mathbf{x}, t)$ is an eigenfunction, whose physical meaning will become clear later.

The operation $\hat{p}_i^2$ is obtained by applying $\hat{p}_i$ twice in succession: $\hat{p}_x^2 = \hat{p}_x \hat{p}_x$. We assume for now that the magnetic vector potential $\mathbf{A}$ is zero. The more general case with nonzero $\mathbf{A}$ will be considered later.

Equating the two expressions (3.8) and (3.10) for the Hamiltonian operator, we can write down the resulting operator equation as

$$i\hbar \frac{\partial}{\partial t} \psi(\mathbf{x}, t) = \left[-\frac{\hbar^2}{2m} \left(\frac{\partial^2}{\partial x^2} + \frac{\partial^2}{\partial y^2} + \frac{\partial^2}{\partial z^2} \right) + q\,\phi(\mathbf{x}) \right] \psi(\mathbf{x}, t). \tag{3.11}$$

The eigenfunction $\psi(\mathbf{x}, t)$ depends on the three spatial coordinates $\mathbf{x} = (x, y, z)$ and the time t. In the following we will make use of the ∇ notation, where, by definition

$$\nabla^2 \psi(\mathbf{x}, t) = \nabla \cdot \nabla \psi(\mathbf{x}, t) = \left(\frac{\partial^2}{\partial x^2} + \frac{\partial^2}{\partial y^2} + \frac{\partial^2}{\partial z^2} \right) \psi(\mathbf{x}, t). \tag{3.12}$$

We therefore write (3.12) as

$$i\hbar \frac{\partial}{\partial t} \psi(\mathbf{x}, t) = -\frac{\hbar^2}{2m} \nabla^2 \psi(\mathbf{x}, t) + q\,\phi(\mathbf{x})\,\psi(\mathbf{x}, t). \qquad (3.13)$$

This is known as the time-dependent Schrödinger equation. It is a linear partial differential equation of second order in $\mathbf{x}$, and first order in t. It applies to general curvilinear coordinates as well as Cartesian coordinates, where one substitutes the appropriate form for the Laplacian operator ∇^2. It can be solved in principle for the eigenfunction $\psi(\mathbf{x}, t)$, given the explicit form for ϕ and appropriate boundary conditions.

We now investigate the physical meaning of the eigenfunction $\psi(\mathbf{x}, t)$. We *assume* that ψ can be written in the separable form

$$\psi(\mathbf{x}, t) = u(\mathbf{x})\,\tau(t), \qquad (3.14)$$

where u is a function only of $\mathbf{x}$ and τ is a function only of t. The function u is not to be confused with the complex transverse particle position in the earlier description of classical geometrical optics. Substituting above, and dividing through by $u\tau$, we obtain

$$i\hbar \frac{1}{\tau(t)} \frac{d}{dt} \tau(t) = \frac{1}{u(\mathbf{x})} \left[-\frac{\hbar^2}{2m} \nabla^2 u(\mathbf{x}) + q\,\phi(\mathbf{x}) \right] = H. \qquad (3.15)$$

We notice that the left side depends only on t, while the middle depends only on $\mathbf{x}$. This is only true in the case where ϕ is independent of time, which we assume for now to be the case. This separation of variables can only hold for all $\mathbf{x}$ and t if both sides are equal to an arbitrary constant, which we call H. The physical meaning of H will become apparent in the following, but for now it is just an arbitrary constant. Substituting, we obtain two separate, decoupled equations given by

$$\frac{d}{dt} \tau(t) + \frac{iH}{\hbar} \tau(t) = 0$$

$$\nabla^2 u(\mathbf{x}) + \frac{2m}{\hbar^2} \left[H - q\phi(\mathbf{x}) \right] u(\mathbf{x}) = 0. \qquad (3.16)$$

The first equation is integrated immediately to give

$$\tau(t) = \tau(0)\, e^{-iHt/\hbar}, \tag{3.17}$$

which the reader can verify by direct substitution. Without loss of generality, we can assume an initial condition $\tau(0) = 1$. The solution for $u(\mathbf{x})$ depends on the particular form for the electrostatic potential $\phi(\mathbf{x})$, which we leave unspecified and general for now. It follows from (3.14) and (3.17) that

$$\psi(\mathbf{x}, t) = u(\mathbf{x})\, e^{-iHt/\hbar}. \tag{3.18}$$

The physical significance of the constant H becomes apparent if we apply the Hamiltonian operator (3.46) to this form for the eigenfunction $\psi(\mathbf{x}, t)$. This is

$$\hat{H}\psi(\mathbf{x}, t) = i\hbar \frac{\partial}{\partial t}\psi(\mathbf{x}, t) = H\psi(\mathbf{x}, t). \tag{3.19}$$

It is immediately apparent from the first postulate above that the constant H is the eigenvalue corresponding to the Hamiltonian operator $\hat{H}$. In the present case, where we assume the potential $\phi(\mathbf{x})$ has no explicit time dependence, the constant H is the eigenvalue representing the conserved total energy.

Depending on the specific boundary conditions, yet to be specified for the particular problem at hand, the Schrödinger equation (3.13) is satisfied only for certain specific values of $u(\mathbf{x})$ and H. We label these $u_j(\mathbf{x})$ and H_j, respectively, where the subscript j is only a label, with integral values assigned for bookkeeping purposes. According to the second postulate above, a single measurement of the total energy H must yield one and only one of the possible values of H_j. The presence of the subscript helps to remind one that the eigenvalues H_j and the dynamical variable corresponding to the classical Hamiltonian function H are two distinct quantities which are related to one another by the formalism just described. Based on this, we can define a set of eigenfunctions which describe the complete behavior of the particle in space and time. This is

$$\psi_j(\mathbf{x}, t) = u_j(\mathbf{x})\, e^{iH_j t/\hbar}, \tag{3.20}$$

which satisfies the time-dependent Schrödinger equation (3.13). The eigenfunction ψ_j oscillates with an angular frequency ω_j defined by

$$H_j = \hbar \omega_j. \qquad (3.21)$$

The angular frequency is constant for a given constant energy eigenvalue H_j, regardless of the form of the electrostatic potential $\phi(\mathbf{x})$, assuming the potential has no explicit time dependence.

We will now proceed to derive two general mathematical properties of u_j and H_j, which will greatly simplify the discussion to follow. From (3.15) we can write

$$-\frac{\hbar^2}{2m} \nabla^2 u_i(\mathbf{x}) + q\,\phi(\mathbf{x})\,u_i(\mathbf{x}) = H_i\, u_i(\mathbf{x})$$

$$-\frac{\hbar^2}{2m} \nabla^2 \bar{u}_j(\mathbf{x}) + q\,\phi(\mathbf{x})\,\bar{u}_j(\mathbf{x}) = \bar{H}_j\, \bar{u}_j(\mathbf{x}) \qquad (3.22)$$

for two different values of the indices i and j, where we have taken the complex conjugate of both sides in the second equation. Multiplying the first equation by $\bar{u}_j$, multiplying the second equation by u_i, and subtracting the second equation from the first, we find

$$-\frac{\hbar^2}{2m}\left[\bar{u}_j(\mathbf{x})\,\nabla^2 u_i(\mathbf{x}) - u_i(\mathbf{x})\,\nabla^2 \bar{u}_j(\mathbf{x}) \right] = (H_i - \bar{H}_j)\,\bar{u}_j(\mathbf{x})\,u_i(\mathbf{x}),$$
$$(3.23)$$

where the potential energy term vanishes, assuming $\phi(\mathbf{x})$ is real.

At this point we specify boundary conditions on $u_i(\mathbf{x})$. We assume that (3.15) is valid only within a cubic volume of side L, where L is arbitrary. Recalling that $\mathbf{x} = (x, y, z)$ in Cartesian coordinates, we further assume that $u_i(x, y, z)$ satisfies the boundary condition

$$u_i(x + L,\, y + L,\, z + L) = u_i(x, y, z). \qquad (3.24)$$

Mathematically, this represents the periodic extension of the wave function over all of space. This is therefore called a *periodic boundary condition*. There is no loss of generality in this assumption, because of the arbitrariness of L. Next, we integrate (3.23) over the cubic volume. This leads to

$$-\frac{\hbar^2}{2m} \int_V d^3\mathbf{x} \left[\bar{u}_j(\mathbf{x})\, \nabla^2 u_i(\mathbf{x}) - u_i(\mathbf{x})\, \nabla^2 \bar{u}_j(\mathbf{x}) \right]$$
$$= (H_i - \bar{H}_j) \int_V d^3\mathbf{x}\, \bar{u}_j(\mathbf{x})\, u_i(\mathbf{x}), \qquad (3.25)$$

where $V = L^3$ is the cubic volume. Rearranging the left-hand side and applying the divergence theorem, we obtain

$$-\frac{\hbar^2}{2m} \int_V d^3\mathbf{x}\, \nabla \cdot \left[\bar{u}_j(\mathbf{x})\, \nabla u_i(\mathbf{x}) - u_i(\mathbf{x})\, \nabla \bar{u}_j(\mathbf{x}) \right]$$
$$= -\frac{\hbar^2}{2m} \int_S d^2 S \left[\bar{u}_j(\mathbf{x})\, \frac{\partial}{\partial n} u_i(\mathbf{x}) - u_i(\mathbf{x})\, \frac{\partial}{\partial n} \bar{u}_j(\mathbf{x}) \right]$$
$$= 0, \qquad (3.26)$$

where S is the surface of the cube. The integral vanishes because of the periodic boundary condition, together with the fact that the normal derivative is equal and opposite on opposite sides of the cube. We therefore have

$$(H_i - \bar{H}_j) \int_V d^3\mathbf{x}\, \bar{u}_j(\mathbf{x})\, u_i(\mathbf{x}) = 0. \qquad (3.27)$$

In the case $i = j$, we assume that the eigenfunction $u_i(\mathbf{x})$ is normalized so that

$$\int_V d^3\mathbf{x}\, \bar{u}_i(\mathbf{x})\, u_i(\mathbf{x}) = 1. \qquad (3.28)$$

It follows that

$$H_i = \bar{H}_i. \qquad (3.29)$$

Equivalently, all energy eigenvalues H_i must be real-valued. In the case $i \neq j$, and assuming $H_i \neq H_j$, the integral in (3.27) must vanish. We therefore have the general property

$$\int_V d^3\mathbf{x}\, \bar{u}_j(\mathbf{x})\, u_i(\mathbf{x}) = \delta_{ij}, \qquad (3.30)$$

where $\delta_{ij} = 0$ for $i \neq j$, and $\delta_{ij} = 1$ for $i = j$. The integral is performed over the cubic volume. This property is known as *orthonormality* of the eigenfunctions $u_i(\mathbf{x})$. Equations (3.29) and

(3.30) represent two very useful mathematical properties in the discussion to follow.

According to the second postulate above, a single precise measurement of the dynamical variable H must yield one and only one of the eigenvalues H_j. It is logical to enquire what determines which of the eigenvalues it must be, or is likely to be, given a specified experimental condition. We now turn our attention to this question. We *define* a function

$$\Psi(\mathbf{x}, t) = \sum_j a_j \, \psi_j(\mathbf{x}, t), \qquad (3.31)$$

where all eigenfunctions $\psi_j(\mathbf{x}, t)$ are assumed to satisfy the time-dependent Schrödinger equation (3.13), and where the $\{a_j\}$ represent a set of complex constants, whose values have yet to be determined. It is straightforward to show by direct substitution that $\Psi(\mathbf{x}, t)$ satisfies the Schrödinger equation as well, namely,

$$-\frac{\hbar^2}{2m}\nabla^2\Psi(\mathbf{x}, t) + q\,\phi(\mathbf{x})\,\Psi(\mathbf{x}, t) = i\hbar\,\frac{\partial}{\partial t}\Psi(\mathbf{x}, t). \qquad (3.32)$$

The proof of this is left as a problem at the end of this section.

We now enquire into the physical interpretation of the function $\Psi(\mathbf{x}, t)$. Writing out the explicit form of the time-dependent Schrödinger equation, together with its complex conjugate equation, we find

$$i\hbar\,\frac{\partial}{\partial t}\Psi(\mathbf{x}, t) = -\frac{\hbar^2}{2m}\nabla^2\Psi(\mathbf{x}, t) + q\,\phi(\mathbf{x})\,\Psi(\mathbf{x}, t)$$

$$-i\hbar\,\frac{\partial}{\partial t}\bar{\Psi}(\mathbf{x}, t) = -\frac{\hbar^2}{2m}\nabla^2\bar{\Psi}(\mathbf{x}, t) + q\,\phi(\mathbf{x})\,\bar{\Psi}(\mathbf{x}, t). \quad (3.33)$$

Multiplying the first of these by $\bar{\Psi}$, multiplying the second by Ψ, and subtracting the second equation from the first, we find

$$i\hbar\left(\Psi\,\frac{\partial}{\partial t}\bar{\Psi} + \Psi\,\frac{\partial}{\partial t}\bar{\Psi}\right) = -\frac{\hbar^2}{2m}\left(\bar{\Psi}\,\nabla^2\Psi - \Psi\,\nabla^2\bar{\Psi}\right). \qquad (3.34)$$

We can rewrite this as

$$\nabla \cdot \left[\frac{i\hbar}{2m} \left(\Psi \nabla \bar{\Psi} - \bar{\Psi} \nabla \Psi \right) \right] + \frac{\partial}{\partial t} \left(\bar{\Psi} \Psi \right) = 0. \qquad (3.35)$$

This has a clear physical interpretation. We define a three-vector quantity $\mathbf{J}(\mathbf{x}, t)$ as

$$\mathbf{J}(\mathbf{x}, t) = \frac{i\hbar}{2m} \left(\Psi \nabla \bar{\Psi} - \bar{\Psi} \nabla \Psi \right), \qquad (3.36)$$

and a function $P(\mathbf{x}, t)$ as

$$P(\mathbf{x}, t) = \bar{\Psi}(\mathbf{x}, t) \cdot \Psi(\mathbf{x}, t) = |\Psi(\mathbf{x}, t)|^2, \qquad (3.37)$$

which is positive-definite. The above equation can be rewritten as

$$\nabla \cdot \mathbf{J} + \frac{\partial}{\partial t} P = 0. \qquad (3.38)$$

We immediately recognize this as a conservation equation, in analogy with fluid flow, where $P(\mathbf{x}, t)$ represents a density, and $\mathbf{J}(\mathbf{x}, t)$ represents a flux.

Based on these mathematical arguments, we identify the quantity $P(\mathbf{x}, t)\, d^3\mathbf{x}$ as the probability that a single precise measurement of the particle position will find the particle in a volume element $d^3\mathbf{x}$ about the position $\mathbf{x}$ at time t. We therefore call the quantity $P(\mathbf{x}, t)$ a *probability density*. As a consistency check, we form the integral over the cubic volume V,

$$\int_V d^3\mathbf{x}\, \nabla \cdot \mathbf{J} + \frac{\partial}{\partial t} \int_V d^3\mathbf{x}\, P(\mathbf{x}, t) = 0. \qquad (3.39)$$

Using the divergence theorem, the leftmost term can be rewritten as

$$\int_V d^3\mathbf{x}\, \nabla \cdot \mathbf{J} = \int_S \mathbf{J} \cdot d\mathbf{S}, \qquad (3.40)$$

where the surface S surrounds the volume V. The integral over the surface S vanishes, owing to the periodic boundary condition, and

the fact that the normal derivative is equal and opposite on opposite sides of the cubic volume. It follows that the time derivative
in (3.39) vanishes. We can therefore write

$$\int_V d^3\mathbf{x}\, P(\mathbf{x}, t) = \int_V d^3\mathbf{x}\, |\Psi(\mathbf{x}, t)|^2 = 1, \tag{3.41}$$

where we have made use of the fact that $\Psi(\mathbf{x}, t)$ can be multiplied by an arbitrary constant, and still satisfy the time-dependent
Schrödinger equation. According to the probability hypothesis,
this is physically equivalent to the fact that the particle is certain to be found somewhere within the volume V.

Substituting,

$$\begin{aligned}
\int_V d^3\mathbf{x}\, |\Psi(\mathbf{x}, t)|^2 &= \int_V d^3\mathbf{x} \left[\sum_i \bar{a}_i\, \bar{\psi}_i(\mathbf{x}, t)\right] \left[\sum_j a_j\, \psi_j(\mathbf{x}, t)\right] \\
&= \sum_{i,j} \bar{a}_i\, a_j\, e^{-i(H_j - H_i)t/\hbar} \int_V d^3\mathbf{x}\, \bar{u}_i(\mathbf{x})\, u_j(\mathbf{x}).
\end{aligned} \tag{3.42}$$

Equivalently,

$$\int_V d^3\mathbf{x}\, |\Psi(\mathbf{x}, t)|^2 = \sum_j |a_j|^2 = 1. \tag{3.43}$$

At this point, we invoke a third key postulate, due originally to
Born [9], [10]:

*The quantity $|a_j|^2$ represents the probability that any single precise
measurement of the total energy will yield the energy eigenvalue
H_j.*

From (3.43) the individual probabilities sum to unity as required.
The set of $\{a_j\}$ are referred to as the *state vector*, and the function
$\Psi(\mathbf{x}, t)$ is called the *state function*.

Based on this probability interpretation, we now define the *expectation value* $\langle H \rangle$ of the total energy at time t as

$$\langle H \rangle = \int_V d^3\mathbf{x}\, \bar{\Psi}(\mathbf{x}, t) \left[\hat{H}\, \Psi(\mathbf{x}, t)\right]. \tag{3.44}$$

Substituting the definition (3.31) for the state function $\Psi(\mathbf{x}, t)$, it is straightforward to show that

$$\langle H \rangle = \sum_j |a_j|^2 H_j \tag{3.45}$$

as expected.

The state function $\Psi(\mathbf{x}, t)$ can be written in terms of the set $\{a_j\}$ as

$$\Psi(\mathbf{x}, t) = \sum_j a_j u_j(\mathbf{x}) e^{-iH_j t/\hbar}. \tag{3.46}$$

Multiplying both sides from the left by $\bar{u}_i$ and integrating over the volume V, we find

$$\int_V d^3\mathbf{x}\, \bar{u}_i(\mathbf{x})\, \Psi(\mathbf{x}, t) = \sum_j a_j\, e^{-iH_j t/\hbar} \int_V d^3\mathbf{x}\, \bar{u}_i(\mathbf{x})\, u_j(\mathbf{x}). \tag{3.47}$$

Making use of the orthonormality of the u_j, this is just

$$a_i = e^{iH_i t/\hbar} \int_V d^3\mathbf{x}\, \bar{u}_i(\mathbf{x})\, \Psi(\mathbf{x}, t). \tag{3.48}$$

Given the state function $\Psi(\mathbf{x}, t)$, we have thus calculated the coefficients a_i of the state vector. The equations (3.48) and (3.46) are therefore the inverse of one another.

We now turn our attention to the relationship between theory and measurement. We will do this in the context of a beam of charged particles, although the thought process applies to other quantum mechanical systems as well. The foregoing analysis applies to a single particle. All relevant information is contained in the state function $\Psi(\mathbf{x}, t)$ and the state vector $\{a_j\}$. The absolute square of the state function is the probability density that a single precise measurement will find the particle at position x at time t. The absolute square of any coefficient a_j is the probability that a single precise measurement of the energy will yield the eigenvalue H_j. We can consider the particle to *exist* in a particular state, as completely specified by these quantities.

One would naturally ask how the particle comes to exist in one particular state, out of a multiplicity of possible states. The answer is determined by the initial experimental condition by which the state of the particle is *prepared.*

For example, we might have a case where the particle has passed through an energy filter, so that the magnitude of the momentum is selected to have one particular known value. Downstream from this, the particle might be normally incident on a pair of slits in an otherwise opaque screen, such that it must pass through one of the slits. However it is not known which slit the particle passes through. This initial condition defines the state function on the front side of the screen as a constant which is independent of transverse position. Furthermore, the state vector roughly has one particular coefficient a_j equal to one, with all other coefficients equal to zero. We will see in the following sections that this is a monoenergetic plane wave, and the experiment is the familiar two-slit experiment. This represents a particular preparation of the state, with this preparation being achieved in a controllable manner experimentally. With this as an initial condition, the state function and state vector then propagate through space and time, consistent with the dynamical equation of motion (3.32).

In this example, one might allow the particle to impinge on a phosphor screen placed further downstream. A flash of light is emitted at the landing position of the particle. The transverse position of the particle is thus measured with high precision. The position of a single particle at a single time does not represent a great deal of useful information, however. It is much more useful to measure the macroscopic properties of a beam of particles. These properties include the current, the distribution of intensity as a function of transverse position, the distribution of intensity as a function of angle, and the distribution of kinetic energies. In the present example the first two properties are measured. One can envision an alternative measurement in which the phosphor screen is replaced by a movable detector or a spectrometer. These would allow measurement of the intensity as a function of angle or the distribution of energies. Measurement of these macroscopic beam properties is conceptually equivalent to repeating the single-particle experiment many times, once for each particle in the beam. The preparation of the single-particle state is assumed to be the same for each particle. This is the conceptual connection between theory and measurement.

In order to better appreciate the physical significance of the theory, we now explore an important and useful special case, namely, a particle moving in a field-free space. This is the subject of the next section.

Problems

1. Construct explicit expressions for the operators representing the three Cartesian components of angular momentum.

2. By definition, a linear operator $\hat{C}$ satisfies

$$\hat{C}\left(c_1\,\varphi_1 + c_2\,\varphi_2\right) = c_1\hat{C}\,\varphi_1 + c_2\hat{C}\,\varphi_2, \tag{3.49}$$

where c_1 and c_2 are *any* two complex constants. Examples of linear operations include multiplication by a constant and differentiation, to name just two. Prove that all of the operators discussed in this section are linear.

3. Prove that the operator $\hat{C}\,\hat{C}$ is linear if $\hat{C}$ is linear.

4. The *commutator* of two operators $\hat{C}_1$ and $\hat{C}_2$ is defined as

$$[\hat{C}_1, \hat{C}_2] \equiv \hat{C}_1\,\hat{C}_2 - \hat{C}_2\,\hat{C}_1. \tag{3.50}$$

(a) Write down an explicit expression for $[\hat{x}, \hat{P}_x]$, where $\hat{x}$ is the operator for the x-coordinate, and $\hat{P}_x$ is the operator for the x-component of the canonical momentum.

(b) Write down an explicit expression for $[\hat{x}, \hat{P}_y]$, where $\hat{x}$ is the operator for the x-coordinate, and $\hat{P}_y$ is the operator for the y-component of the canonical momentum.

5. Prove that the state function $\Psi(\mathbf{x}, t)$ defined by (3.31) satisfies the time-dependent Schrödinger equation (3.32).

3.1.2 Particle motion in a field-free space

The special case of a particle in field-free space is represented by $\phi(\mathbf{x}, t) = 0$ and $\mathbf{A}(\mathbf{x}, t) = 0$, where ϕ and $\mathbf{A}$ represent the electrostatic scalar potential, and the magnetic vector potential, respectively. The spatial part of the Schrödinger equation in (3.16) reduces to

$$\nabla^2 u(\mathbf{x}) + \frac{2mH}{\hbar^2}\, u(\mathbf{x}) = 0, \tag{3.51}$$

where H is the conserved total energy. Equivalently,

$$\nabla^2\, u(\mathbf{x}) + k^2\, u(\mathbf{x}) = 0, \tag{3.52}$$

where we have defined a constant k by

$$k^2 = \frac{2mH}{\hbar^2}. \tag{3.53}$$

We propose to integrate this using separation of variables, similar to the previous section. We *assume* that the eigenfunction $u(\mathbf{x})$ can be expressed in Cartesian coordinates in separable form as

$$u(x, y, z) = X(x)\, Y(y)\, Z(z). \tag{3.54}$$

Substituting, and dividing through by XYZ, this leads to

$$\frac{X''(x)}{X(x)} + \frac{Y''(y)}{Y(y)} + \frac{Z''(z)}{Z(z)} + k^2 = 0. \tag{3.55}$$

This, in turn, leads to three independent equations

$$\begin{aligned} X''(x) + k_x^2\, X(x) &= 0 \\ Y''(y) + k_y^2\, Y(y) &= 0 \\ Z''(z) + k_z^2\, Z(z) &= 0, \end{aligned} \tag{3.56}$$

where we have defined three separate constants k_x, k_y, and k_z which obey

$$k^2 = k_x^2 + k_y^2 + k_z^2. \tag{3.57}$$

Taking the first equation in (3.56), this has two independent solutions given by

$$X(x) = e^{\pm i k_x x}, \tag{3.58}$$

which the reader can verify by direct substitution. We immediately recognize a problem, in that the integral of $|X(x)|^2$ over the range of x from $-\infty$ to ∞ is infinite. This is inconsistent with the probabilistic interpretation of eigenfunctions.

Our analysis is incomplete to this point, however, because we have yet to specify the boundary conditions. To resolve this, we first impose the arbitrary condition that $X(x)$ is defined only over the range $-L/2 \le x \le L/2$. We further impose the periodic boundary condition by assuming

$$X(x + L) = X(x), \tag{3.59}$$

which we are completely at liberty to do, without loss of generality. Substituting, this yields

$$e^{\pm i k_x L} = 1. \tag{3.60}$$

This, in turn, requires that the constant k_x take on discrete values

$$k_x = \frac{2\pi n_x}{L}, \qquad n_x = 0, \pm 1, \pm 2, \ldots. \tag{3.61}$$

In addition, the solution can be multiplied by an arbitrary constant, without affecting its validity. This gives

$$X(x) = \frac{1}{\sqrt{L}}\, e^{i k_x x}. \tag{3.62}$$

It follows that

$$\int_{-L/2}^{L/2} dx\, |X(x)|^2 = 1, \qquad (3.63)$$

thus satisfying the normalization condition, required for probability.

Repeating this for the y- and z-equations, we obtain

$$
\begin{aligned}
u_{\mathbf{k}}(\mathbf{x}) &= X(x)\,Y(y)\,Z(z) \\
&= \frac{1}{\sqrt{L^3}}\, e^{i\,(k_x\, x + k_y\, y + k_z\, z)} \\
&= \frac{1}{\sqrt{V}}\, e^{i\mathbf{k}\cdot\mathbf{x}}, \qquad (3.64)
\end{aligned}
$$

where $V = L^3$ is the volume of the cube. We have adopted the vector notation $\mathbf{k} = (k_x, k_y, k_z)$. The components k_x, k_y, and k_z take on the discrete values

$$ k_x = \frac{2\pi n_x}{L}, \qquad k_y = \frac{2\pi n_y}{L}, \qquad k_z = \frac{2\pi n_z}{L}, \qquad (3.65) $$

where $n_j = 0, \pm 1, \pm 2, \dots$. The vector $\mathbf{k}$ is called the *wave vector*. It is straightforward to show that

$$\int_V d^3 x\, \bar{u}_{\mathbf{k}}(\mathbf{x})\, u_{\mathbf{k}'}(\mathbf{x}) = \delta_{\mathbf{k}\mathbf{k}'}, \qquad (3.66)$$

where the integral is performed over the cubic volume V. The eigenfunctions $u_{\mathbf{k}}(\mathbf{x})$ are plane waves, each with a unique wave vector $\mathbf{k}$.

Each set of n_x, n_y, and n_z represents a distinct state with energy given by

$$ H_k = \frac{\hbar^2}{2m}\left(k_x^2 + k_y^2 + k_z^2\right) = \frac{\hbar^2}{2mL^2}\left(n_x^2 + n_y^2 + n_z^2\right). \qquad (3.67) $$

The interval L can be chosen to be arbitrarily large. As L is increased, the energy values become more closely spaced. In the limit $L \to \infty$, the energy levels approach a continuum. It does *not* follow that the energy eigenvales become small, since the integers n_x,

n_y, and n_z can take on arbitrarily large values.

The eigenvalues (k_x, k_y, k_z) form an infinite cubic lattice of equally spaced points in k-space. Each lattice point can be regarded as occupying a cubic volume element of $(2\pi/L)^3$ around the lattice point in k-space. Based on this, the number of states per unit volume in k-space is given by

$$\frac{dN}{dV_k} = \frac{V}{(2\pi)^3},\qquad (3.68)$$

where, again, $V = L^3$. A unique wave vector $\mathbf{k}$ exists for each lattice point with components given by

$$\mathbf{k} = (k_x, k_y, k_z).\qquad (3.69)$$

From (3.67), the discrete energy eigenvalue associated with each lattice point is

$$H_k = \frac{\hbar^2 k^2}{2m},\qquad (3.70)$$

where k is the magnitude of the wave vector $\mathbf{k}$. It follows that the surfaces of constant energy in k-space are spheres of radius k about the origin $\mathbf{k} = (0,0,0)$. A small energy interval is therefore represented by a spherical shell of volume dV_k and thickness dk where

$$dV_k = 4\pi\, k^2 dk.\qquad (3.71)$$

We can calculate the number of states per unit energy interval. This is given using the chain rule for derivatives as

$$\frac{dN}{dH_k} = \frac{dN}{dV_k}\frac{dV_k}{dk}\frac{dk}{dH_k}.\qquad (3.72)$$

It is straightforward to show from (3.68, 3.70, 3.71, 3.72) that

$$\frac{dN}{dH_k} = \frac{4\pi V}{h^3}\sqrt{2\, m^3 H_k}.\qquad (3.73)$$

This quantity will turn out to be very useful later on. In words, the density of energy states is proportional to the square root of the energy. This calculation shows the simplification which results from regarding the states in $k-$space.

Separately, it is interesting to see what happens when we apply the operator for the canonical momentum to the energy eigenfunctions. This gives

$$-i\hbar \, \nabla u_{\mathbf{k}}(\mathbf{x}) = \hbar \mathbf{k} \, u_{\mathbf{k}}(\mathbf{x}). \tag{3.74}$$

Evidently, the energy eigenfunctions $u_{\mathbf{k}}(\mathbf{x})$ are also the eigenfunctions of the canonical mometum operator, with eigenvalues $\hbar \mathbf{k}$. Having assumed that the magnetic vector potential $\mathbf{A}$ is zero, we therefore identify $\hbar \mathbf{k}$ with the kinetic momentum. This momentum is proportional to the gradient of $u_{\mathbf{k}}(\mathbf{x})$. It follows that the vector $\mathbf{k}$ is perpendicular to the surfaces $u = const$, and therefore points in the direction of wave propagation, as expected. The wavelength is given by

$$\lambda = \frac{2\pi}{k}. \tag{3.75}$$

This is the de Broglie wavelength, given by $\lambda = h/p$, where p is the momentum.

Including the time dependence (3.18), we have

$$\psi_{\mathbf{k}}(\mathbf{x}, t) = u_{\mathbf{k}}(\mathbf{x}) \, e^{-iH_k t/\hbar} = \frac{1}{\sqrt{V}} \, e^{i(\mathbf{k}\cdot\mathbf{x} - \omega_k t)}, \tag{3.76}$$

where $H_k = \hbar \omega_k$. This is a traveling plane wave, propagating in the direction $\mathbf{k}$. From (3.57) the wave vector $\mathbf{k}$ and the angular frequency ω_k are related by the energy-momentum equation as

$$\omega_k = \frac{\hbar k^2}{2m}, \tag{3.77}$$

where $k^2 = |\mathbf{k}|^2 = \mathbf{k} \cdot \mathbf{k}$. This is called the *dispersion relation*. The wave propagates with phase velocity v_p given by

$$v_p = \frac{\omega_k}{k} = \frac{\hbar k}{2m}. \tag{3.78}$$

Evidently, states with higher k (shorter wavelength) propagate faster than states with lower k.

Following the analysis of the preceding section, we define a state function

$$
\begin{aligned}
\Psi(\mathbf{x}, t) &= \sum_{\mathbf{k}} a_{\mathbf{k}}\, \psi_{\mathbf{k}}(\mathbf{x}, t) \\
&= \sum_{\mathbf{k}} a_{\mathbf{k}}\, u_{\mathbf{k}}(\mathbf{x})\, e^{-iH_{k}t/\hbar} \\
&= \frac{1}{\sqrt{V}} \sum_{\mathbf{k}} a_{\mathbf{k}}\, e^{i(\mathbf{k}\cdot\mathbf{x} - \omega_{k}t)},
\end{aligned}
\tag{3.79}
$$

where the summation over $\mathbf{k}$ represents a summation over all possible values of (n_x, n_y, n_z). As previously, the probability that a single precise measurement of the total energy yields a specific value $H_{\mathbf{k}}$ is given by $|a_{\mathbf{k}}|^2$. Following the procedure of the earlier analysis, we find

$$
a_{\mathbf{k}} = e^{i\omega_{k}t}\, \frac{1}{\sqrt{V}} \int_{V} d^3\mathbf{x}\, \Psi(\mathbf{x}, t)\, e^{-i\mathbf{k}\cdot\mathbf{x}},
\tag{3.80}
$$

where the integral is over the cubic volume V.

Next we *define* a function $\Phi(\mathbf{k}, t)$ which satisfies

$$
\frac{a_{\mathbf{k}}}{\sqrt{V}} = \frac{1}{(2\pi)^{3/2}} \frac{dV_{\mathbf{k}}}{dN}\, \Phi(\mathbf{k}, t).
\tag{3.81}
$$

The reason for this precise definition will become clear shortly. Substituting this into (3.79) and making use of (3.68), the state function is

$$
\Psi(\mathbf{x}, t) = \frac{1}{(2\pi)^{3/2}} \sum_{\mathbf{k}} \frac{dV_{\mathbf{k}}}{dN}\, \Phi(\mathbf{k}, t)\, e^{i(\mathbf{k}\cdot\mathbf{x} - i\omega_{k}t)}.
\tag{3.82}
$$

We now consider the limiting case where the cubic volume V is taken to be very large. According to the preceding arguments, the lattice of eigenstates in $\mathbf{k}$-space becomes very dense. In this case the sum can be represented by the integral

$$
\Psi(\mathbf{x}, t) = \frac{1}{(2\pi)^{3/2}} \int d^3\mathbf{k}\, \Phi(\mathbf{k}, t)\, e^{i(\mathbf{k}\cdot\mathbf{x} - \omega_{k}t)},
\tag{3.83}
$$

where $d^3\mathbf{k}$ is the volume element in $\mathbf{k}$-space corresponding to one state ($dN = 1$). Separately from (3.68, 3.81),

$$\frac{a_\mathbf{k}}{\sqrt{V}} = \frac{(2\pi)^{3/2}}{V} \Phi(\mathbf{k}, t) = e^{i\omega_k t} \frac{1}{V} \int_V d^3\mathbf{x}\, \Psi(\mathbf{x}, t)\, e^{-i\mathbf{k}\cdot\mathbf{x}}, \qquad (3.84)$$

where the integral is over the cubic volume V. Again taking the limit of V very large, this becomes equivalent to

$$\Phi(\mathbf{k}, t) = \frac{1}{(2\pi)^{3/2}} \int d^3\mathbf{x}\, \Psi(\mathbf{x}, t)\, e^{-i(\mathbf{k}\cdot\mathbf{x}-\omega_k t)}, \qquad (3.85)$$

where the integral is now over all space. In order for this integral to converge, it is necessary that the state function falls to zero for very large $\mathbf{x}$. This is equivalent to saying that the particle is *localized* over some finite region of space.

The physical significance of $\Phi(\mathbf{k}, t)$ can be appreciated by forming the integral

$$\int d^3\mathbf{k}\, |\Phi(\mathbf{k}, t)|^2 = \int d^3\mathbf{k} \left[\frac{1}{(2\pi)^{3/2}} e^{i\omega_k t} \int d^3\mathbf{x}\, \Psi(\mathbf{x}, t)\, e^{-i\mathbf{k}\cdot\mathbf{x}} \right]$$
$$\cdot \left[\frac{1}{(2\pi)^{3/2}} e^{-i\omega_k t} \int d^3\mathbf{x}'\, \bar{\Psi}(\mathbf{x}', t)\, e^{i\mathbf{k}\cdot\mathbf{x}'} \right].$$
$$(3.86)$$

Rearranging the order of integrations, this is equivalent to

$$\int d^3\mathbf{k}\, |\Phi(\mathbf{k}, t)|^2 = \int d^3\mathbf{x}\, \Psi(\mathbf{x}, t) \int d^3\mathbf{x}'\, \bar{\Psi}(\mathbf{x}', t)$$
$$\cdot \left[\frac{1}{(2\pi)^3} \int d^3\mathbf{k}\, e^{-i\mathbf{k}\cdot(\mathbf{x}-\mathbf{x}')} \right]. \qquad (3.87)$$

We recognize the quantity in square brackets as the Dirac delta function, namely,

$$\delta(\mathbf{x} - \mathbf{x}') = \frac{1}{(2\pi)^3} \int d^3\mathbf{k}\, e^{-i\mathbf{k}\cdot(\mathbf{x}-\mathbf{x}')}. \qquad (3.88)$$

Making use of the property of the delta function, this leads immediately to

$$\int d^3k\,|\Phi(\mathbf{k},t)|^2 = \int d^3x\,|\Psi(\mathbf{x},t)|^2 = 1. \tag{3.89}$$

From this we interpret $|\Phi(\mathbf{k},t)|^2$ as the probability density in $\mathbf{k}$-
space, and $\Phi(\mathbf{k},t)$ as the state function in $\mathbf{k}$-space. Recalling that
the momentum is $\mathbf{p} = \hbar\mathbf{k}$, it follows that $\Phi(\mathbf{k},t)$ describes the
state in momentum space. From (3.83) and (3.85) we see that the
state functions $\Psi(\mathbf{x},t)$ and $\Phi(\mathbf{k},t)$ are related by a Fourier transform with respect to the spatial variables, but not with respect to
time. The result (3.89) is a general property of Fourier transforms
known as Parseval's theorem.

As a further example of free-particle propagation, we consider
two sources at $x = \pm\infty$, which radiate in phase with each other.
By the earlier analysis, this gives rise to two individual free-particle
eigenstates, with normalized eigenfunctions given respectively by

$$\psi_+(x,t) \;=\; \frac{1}{\sqrt{L}}\,\exp[\,i(+kx - \omega t)\,]$$

$$\psi_-(x,t) \;=\; \frac{1}{\sqrt{L}}\,\exp[\,i(-kx - \omega t)\,], \tag{3.90}$$

where $\hbar k$ is the momentum and $\hbar\omega$ is the energy. The problem is
defined on the interval $-L/2 \le x \le +L/2$. Consistent with the
earlier analysis, we assume periodic boundary conditions, where k
takes on discrete values $k_n = 2\pi n/L$, which approach a continuum
as L approaches infinity. These two plane waves propagate in opposite directions. The combination of these waves is represented
by the superposition state

$$\Psi(x,t) = \frac{1}{\sqrt{2}}\,(\psi_+ + \psi_-). \tag{3.91}$$

Substituting, this is

$$\Psi(x,t) = \sqrt{\frac{2}{L}}\,\cos(k_n x)\,e^{-i\omega_n t}, \qquad\qquad n = 0, \pm 1, \pm 2, \ldots. \tag{3.92}$$

The resulting intensity is

$$|\Psi(x,t)|^2 = \frac{2}{L}\cos^2(k_n x).$$ (3.93)

We notice immediately that the time has dropped out, thus forming a standing wave. This satisfies the normalization condition

$$\int_{-L/2}^{L/2} dx \, |\Psi(x,t)|^2 = 1$$ (3.94)

for all wave numbers $k_n = 2\pi n/L$. The intensity is plotted in Figure 3.1 for the case $n = 2$. The intensity exhibits bright and

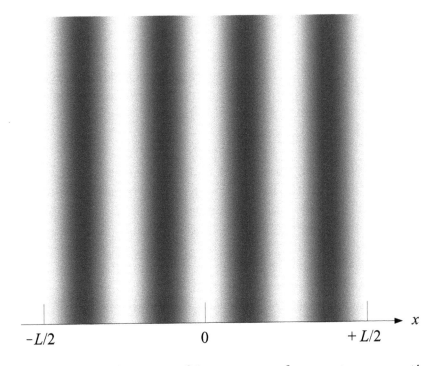

$-L/2$ 0 $+L/2$ x

Figure 3.1: Standing-wave fringe pattern for counterpropagating plane waves.

dark fringes, indicating constructive and destructive interference, respectively. The spatial period of the fringes is inversely proportional to k_n, which can take on a multiplicity of values.

Experimentally the intensity distribution is a property of the two sources, i.e., the way in which the system is prepared. The two sources are said to radiate *coherently*, because they have a definite, constant phase relationship to one another. This situation can be realized experimentally by impinging a parallel beam on a positively charged fiber in otherwise field-free space. The beam is bent toward the fiber on both sides, thus creating a region where the beams from the two sides interact coherently with each other. Such an arrangement is called a *biprism*, and has been demonstrated. We have seen from this example that a single-particle state with just two individual eigenstates populated shows striking and distinctive interference behavior.

Problems

1. Write down explicit expressions for the normalized eigenfunctions, eigenvalues, and dispersion relations for a free particle in one Cartesian dimension, assuming periodic boundary conditions on a spatial interval of length L.

2. Estimate the quantum number n_x for a free electron with energy 1 keV moving in a drift length $L = 1$ m. The correspondence principle states that quantum mechanical particle motion approaches classical behavior in the limit of large quantum numbers.

3. The state function $\Psi(\mathbf{x}, t)$ is said to describe single-particle motion in the *energy* representation, since the eigenvalues of the Hamiltonian operator $\hat{H}$ represent conserved energy. The state function $\Phi(\mathbf{k}, t)$ is said to describe the *momentum* representation. Write down explicit expressions for the eigenfunctions and eigenvalues for a free particle in the momentum representation.

3.1.3 Wave packet propagation and the Heisenberg uncertainty principle

From (3.79) the state function $\Psi(\mathbf{x}, t)$ represents a superposition of indivudual plane waves propagating in space and time. Each plane wave is described by a single eigenfunction $\psi_{\mathbf{k}}(\mathbf{x}, t)$ with well-defined wave vector eigenvalue $\mathbf{k}$ and angular frequency eigenvalue ω_k. All of the individual plane waves interfere with one another to form a wave *packet*. This describes the propagation of a *single* particle. We can gain an intuitive feel for this by considering one spatial dimension. The eigenfunction for a single state k is

$$\psi_k(x, t) = \frac{1}{\sqrt{L}} e^{i(kx - \omega_k t)}. \tag{3.95}$$

Assuming the wave packet consists of individual eigenstates which are close together in energy, there is some central (k_0, ω_0) for which the waves interfere constructively. This is represented by an extremum condition

$$\left[\frac{d}{d\omega} (kx - \omega t) \right]_{k_0, \omega_0} = 0, \tag{3.96}$$

where the derivative is evaluated at (k_0, ω_0). Performing the differentiation, we find

$$\frac{x}{t} = \left[\frac{d\omega}{dk} \right]_{k_0, \omega_0}, \tag{3.97}$$

where the left side is the velocity of propagation. We thus define the *group velocity* as

$$v_g = \left[\frac{d\omega}{dk} \right]_{k_0, \omega_0}. \tag{3.98}$$

Generalizing this to three dimensions, this is

$$\mathbf{v}_g = \left[\nabla_{\mathbf{k}} \omega(\mathbf{k}) \right]_{\mathbf{k}_0, \omega_0}. \tag{3.99}$$

We see from the dispersion relation that

$$\mathbf{v}_g = \frac{\hbar \mathbf{k}_0}{m}. \tag{3.100}$$

Since the numerator is the momentum, we identify the group velocity with the classical particle velocity. In this sense, the motion of the group corresponds to the classical particle motion.

To further illustrate the significance of the state function, we consider a particular example. We assume that the system can be prepared experimentally, so that $\Psi(x,0)$ takes the form

$$\Psi(x,0) = \left[\frac{1}{\sqrt{2\pi\sigma^2}} \exp\left(-\frac{x^2}{2\sigma^2}\right) \right]^{1/2} \qquad (3.101)$$

in one dimension. This is intentionally constructed so that $|\Psi(x,0)|^2$ is a Gaussian distribution, and the integral over $-\infty < x < \infty$ is unity, as required for a probability distribution. This is often referred to as a Gaussian wave packet. The quantity σ is known as the standard deviation, and is a measure of the width of $|\Psi(x,0)|^2$. We therefore define the uncertainty in the x-coordinate as

$$\Delta x = \sigma. \qquad (3.102)$$

Next, we form $\Phi(k)$ in one dimension. This is

$$\Phi(k) = \frac{1}{2\pi} \int_{-\infty}^{\infty} dx\, \Psi(x,0)\, e^{-ikx}. \qquad (3.103)$$

Substituting for $\Psi(x,t)$, we find

$$\Phi(k) = \frac{1}{2\pi} \frac{1}{(2\pi\sigma^2)^{1/4}} \int_{-\infty}^{\infty} e^{-a^2 x^2} e^{-ikx}, \qquad (3.104)$$

where we have defined

$$a^2 = \frac{1}{4\sigma^2}. \qquad (3.105)$$

From tables,

$$\frac{1}{\sqrt{2\pi}} \int_{-\infty}^{\infty} e^{-a^2 x^2} e^{-ikx} = \frac{1}{a\sqrt{2}} \exp\left(-\frac{k^2}{4a^2}\right). \qquad (3.106)$$

This gives

$$|\Phi(k)|^2 = \frac{1}{\sigma a^2 \sqrt{32\pi^3}} \exp\left(-\frac{k^2}{2a^2}\right). \qquad (3.107)$$

The standard deviation is immediately recognizable as

$$\Delta k = a = \frac{1}{2\sigma} = \frac{1}{2\Delta x}. \tag{3.108}$$

Equivalently,

$$\Delta x \, \Delta k = \tfrac{1}{2}. \tag{3.109}$$

Recalling the momentum $p = \hbar k$, it follows that

$$\Delta x \, \Delta p = \tfrac{1}{2}\hbar. \tag{3.110}$$

This is an example of the *Heisenberg uncertainty principle*, which states that one can never know the position and momentum simultaneously to a precision better than this.

Given that the state function $\Psi(\mathbf{x}, t)$ depends on the experimental conditions, we now return to the question of how one goes about preparing a system experimentally. We described a practical approach to preparing a state consisting of a monochromatic plane wave. As a further example, we now discuss the preparation of a state consisting of a wave packet. We consider an electron beam emitted from a thermionic (hot) source, and accelerated through a potential difference ϕ_0. The electrons in the beam have a spread of energies of the order $\Delta H = kT$, where k is Boltzmann's constant, and T is the absolute temperature of the electron source. The energies are distributed about a central value $H_0 = e\phi_0$. From the energy-momentum relation, we have

$$\hbar k = \sqrt{2mH}. \tag{3.111}$$

Taking the differential of both sides, we find a spread of momentum

$$\Delta p = \hbar \Delta k = \sqrt{\frac{m}{2H}}\,\Delta H = \sqrt{\frac{m}{2\,e\,\phi_0}}\,k\,T. \tag{3.112}$$

Regarding the wave packet as Gaussian, and invoking the uncertainty principle, this leads to an uncertainty in the position of the particle along the beam axis given by

$$\Delta x = \frac{\hbar}{2\,\Delta p} = \frac{\hbar}{kT}\sqrt{\frac{e\phi_0}{2m}}, \tag{3.113}$$

where we can regard Δx as the extent of the wave packet.

In summary, the absolute square $|\Psi(\mathbf{x}, t)|^2$ of the state function is the probability density that any single measurement will find a single particle at position $\mathbf{x}$ at time t. As such, this quantity has primary physical significance. The state function, in turn, is a linear superposition of individual eigenfunctions $\psi_j(\mathbf{x}, t)$, with the coefficients a_j determined by the way in which the system is prepared experimentally. The eigenfunctions ψ_j with associated energy eigenvalues ε_j represent solutions to the Hamiltonian operator equation, which, in turn governs the dynamical behavior of the particle. Any single measurement of a single particle must find the particle in one, and only one eigenstate. In particular, a single measurement of the particle energy yields one, and only one eigenvalue H_j. The probability of finding the particle in the jth eigenstate is $|a_j|^2$.

Problems

1. An electron beam is accelerated to an energy of 1.0 KeV. The beam is then made to pass through an energy filter which transmits only electrons with a spread of energies $\Delta E = 0.025$ eV about the mean energy. Estimate the uncertainty in arrival time of a single electron at a point just at the exit from the energy filter.

2. Electrons are emitted from a cathode and accelerated to form a beam. Describe in words the conceptual relationship between the macroscopic beam properties (current, energy, energy spread, path length, and transit time) and the quantum mechanical motion of a single beam electron. Assume the beam electrons do not interact significantly with one another.

3.1.4   The quantum mechanical analog of Fermat's principle for matter waves

Fermat's principle describes the propagation of light, or more generally electromagnetic radiation. It states that light propagates along a path which minimizes the transit time. Mathematically this is equivalent to the statement that every physically allowable ray satisfies the condition that the line integral of the index of refraction n is stationary with respect to infinitesimal variations. This is

$$\delta \int_{x_a}^{x_b} n \, ds = 0. \tag{3.114}$$

The index of refraction n is a property of the medium through which the light propagates. It is defined as

$$n = \frac{c}{v_p}, \tag{3.115}$$

where c is the speed of light in vacuum, and v_p is the phase velocity of propagation in the particular medium. In general n can vary from point to point in the medium, and is therefore a function of position. In vacuum $v_p = c$ and $n = 1$. Since c is a constant, Fermat's principle can be written in the alternative form

$$\delta \int \frac{ds}{v_p} = 0. \tag{3.116}$$

Taking $v_p = ds/dt$, this says physically that the path chosen by the light ray is the one for which the propagation time is an extremum. In fact, the propagation time is a minimum.

According to an analysis by Fermi [27], a quantum mechanical analogy with Fermat's principle exists, which describes propagation of a single particle. A general property of propagating waves says that the phase velocity can be written as $v_p = \nu\lambda$, where ν is the temporal frequency, and λ is the wavelength. In the case where the electromagnetic potentials have no explicit time dependence, the total energy is conserved. The conserved total energy is $H = h\nu$, from which it follows that ν is a constant. In this case Fermat's principle is equivalent to

$$\delta \int \frac{ds}{\lambda} = 0. \tag{3.117}$$

Physically, all hypothetical rays in an infinitesimal neighborhood surrounding the physical ray interfere constructively. In this context, Fermat's principle is fundamentally wave-mechanical.

Separately, the physical trajectory of a classical point particle with mass m obeys the principle of least action,

$$\delta \int p \, ds = 0, \tag{3.118}$$

where we have assumed that the magnetic vector potential $\mathbf{A}$ is zero, and the electrostatic potential $\phi(x)$ has no explicit time dependence. Substituting for the kinetic momentum p, this is equivalent in one dimension to

$$\delta \int \sqrt{2m[H - U(x)]} \, ds = 0, \tag{3.119}$$

where $U(x) = q\,\phi(x)$ is the potential energy, and H is the conserved total energy. The integrand can be regarded as an index of refraction in the mechanical analog of Fermat's principle in classical mechanics. Thus we have two alternative expressions for the index of refraction. They are not equivalent, since one is wave-mechanical, and the other is derived from classical mechanics.

Following Fermi, equations (3.116) and (3.119) guide us to form a working assumption, namely, the phase velocity v_p can be written in the analogous functional form

$$\frac{1}{v_p} = f(\omega) \sqrt{H(\omega) - U(x)}, \tag{3.120}$$

where $f(\omega)$ and $H(\omega)$ are arbitrary functions of the angular frequency ω, yet to be determined. We will investigate the validity of this assumption in the following.

The phase velocity v_p is given quite generally by

$$v_p = \frac{\omega}{k}, \tag{3.121}$$

where k is the wave number given by $k = 2\pi/\lambda$, and ω is the angular frequency given by $\omega = 2\pi\nu$. Substituting above, this gives

$$k = \omega f(\omega) \sqrt{H(\omega) - U(x)}. \tag{3.122}$$

This represents a dispersion formula, relating the wave number k and the angular frequency ω.

To this point we have regarded k and ω to be fixed, with each taking on a single value. In practice, the quantum mechanical state consists of a superposition of multiple eigenstates, with each state characterized by a unique value of k and a unique value of ω. This superposition represents a wave packet, which propagates with a group velocity v_g, given (3.98) in one dimension by

$$v_g = \frac{d\omega}{dk}. \tag{3.123}$$

The derivative is evaluated at central values of k and ω, for which all partial waves associated with the individual eigenstates interfere constructively. From (3.122) and (3.123) this gives

$$\begin{aligned} \frac{1}{v_g} &= \frac{dk}{d\omega} \\ &= \sqrt{H(\omega) - U(x)}\, \frac{d}{d\omega}\left[\omega\, f(\omega)\right] + \omega\, f(\omega)\, \frac{1}{2\sqrt{H(\omega) - U(x)}}\, \frac{dH}{d\omega}. \end{aligned}$$
$$\tag{3.124}$$

At this point we make a further working assumption, namely

$$\frac{d}{d\omega}\left[\omega\, f(\omega)\right] = 0, \tag{3.125}$$

which we will proceed to vindicate later. It follows from this that

$$\omega\, f(\omega) = const. \tag{3.126}$$

According to the correspondence principle, the group velocity v_g tends to the classical particle velocity in the limit of high quantum numbers. The classical kinetic momentum p above is then replaced by $m v_g$ in the limit. Following Fermi, this prompts us to further assume independently that

$$\frac{1}{v_g} = \frac{1}{\sqrt{(2/m)\,[\,H(\omega) - U(x)\,]}}, \tag{3.127}$$

where we have replaced the classical conserved total energy H with the undetermined function $H(\omega)$. Equating the two expressions (3.124) and (3.127) for $1/v_g$ with the condition (3.125), we find that

$$\frac{dH}{d\omega} = \frac{\sqrt{2m}}{\omega\,f(\omega)} = const. \tag{3.128}$$

It can be shown experimentally, by electron diffraction by crystals, for example, that the constant on the far right must be $\hbar$. This leads to

$$H = \hbar\omega, \qquad \omega\,f(\omega) = \frac{\sqrt{2m}}{\hbar}. \tag{3.129}$$

The total energy eigenvalue H is determined to within an arbitrary additive integration constant. The equation on the right vindicates our assumption (3.126) that $\omega\,f(\omega) = const.$ Substituting above, this leads to

$$k = \frac{\sqrt{2m\,[\,\hbar\omega - U(x)\,]}}{\hbar}. \tag{3.130}$$

Equivalently,

$$\hbar\omega = \frac{\hbar^2 k^2}{2m} + U(x). \tag{3.131}$$

We recognize this as the dispersion relation resulting from conservation of total energy, where $\hbar\omega$ is the total energy eigenvalue, and $\hbar k$ is the momentum eigenvalue. Both ω and k are evaluated at the central values which characterize the wave packet or superposition state.

This analysis shows that the quantum mechanics of single-particle propagation can be described in terms of a variational principle and the resulting dispersion relation. This is completely equivalent to the postulate-based approach described earlier. Physically, a close analogy exists with the wave optics of light propagation.

3.2 Particle motion in a general electromagnetic potential

In the preceding sections we have reviewed the conceptual basis of quantum mechanics, as it relates to the motion of a single particle. We are now in a position to include electric and magnetic effects. For the present purpose, we confine our attention to fields which vary slowly in space and time. This permits a more traditional approach, based on Schrödinger theory. To this end, we consider all relevant information about the electric and magnetic effects to be contained in the electrostatic scalar potential $\phi(\mathbf{x}, t)$ and the magnetic vector potential $\mathbf{A}(\mathbf{x}, t)$, respectively, where these potentials are functions of position $\mathbf{x}$ and time t. Together, these potentials form the components of a Lorentz-covariant four-vector (2.5), which we refer to as a general electromagnetic potential. The central problem is to solve for the wave function $\psi(\mathbf{x}, t)$ in the presence of a general electromagnetic potential, where the absolute square of this function is the probability density for finding the particle at position $\mathbf{x}$ and time t.

3.2.1 Path integral approach for the time-dependent wave function

A great deal of physical insight can be gained from the path integral description of quantum mechanics, originally due to Feynman

[28]. The reader is referred to an emended version by Styer of the
original text by Feynman and Hibbs [29] for a detailed, comprehensive, and highly readable description. Our present goal is to
summarize the highlights of the text, with particular application
to the motion of a single charged particle moving in a general electromagnetic potential.

We begin by studying the motion of a particle in one dimension,
where the position x is a function of time t. Classically, the motion
from an initial time t_a to a later time t_b is along that path which
represents an extremum of the action integral S_{ba} given by

$$S_{ba} = \int_{t_a}^{t_b} L(x, v; t)\, dt, \qquad (3.132)$$

where L is the Lagrangian given by (2.9). The beginning point
charcterized by position and time (x_a, t_a), and the end point characterized by position and time (x_b, t_b) are assumed to be fixed.

In a quantum-mechanical description we seek a probability amplitude φ_{ba} for the particle to propagate from an initial position
and time (x_a, t_a) to a final position and time (x_b, t_b). Again we
assume that the end points (x_a, t_a) and (x_b, t_b) are fixed.

At this point we form a key hypothesis, namely, the amplitude
φ_{ba} can be written for a given path of motion as

$$\varphi_{ba} = const \times \exp\left[\frac{i}{\hbar} S_{ba}\right], \qquad (3.133)$$

where S_{ba} is the action integral. An infinite number of possible
paths exist, each path having a distinct value of S_{ba}. This is illus-
trated for a hypothetical system in Figure 3.2. The general rule
in quantum mechanics is that the amplitudes for all alternative
paths must be added to form the resultant amplitude. The absolute square of this resultant amplitude then represents the probability density for finding the system at a given coordinate x. The
amplitudes φ_{ba} for all possible paths must therefore be summed to
form the overall probability amplitude $K(x_b, t_b; x_a, t_a)$. Following

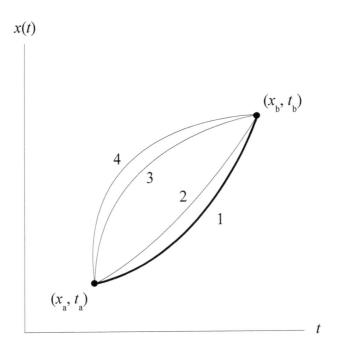

Figure 3.2: Possible paths between two fixed points in one spatial dimension.

Feynman and Hibbs, we will refer to this overall amplitude as a *kernel*, and the summation over all possible paths as a *path integral*.

The absolute square of the kernel represents the probability density P_{ba} for finding the particle at position x_b at time t_b, having started at position x_a at an earlier time t_a. Equivalently,

$$P_{ba} = |K(x_b, t_b; x_a, t_a)|^2. \tag{3.134}$$

For a charged particle optics system of macroscopic dimensions, the action integral S_{ba} is very large compared to $\hbar$. A small variation in path therefore results in a large variation in the action integral divided by $\hbar$, or equivalently in the phase of φ_{ba}. The classical path of motion is labeled 1 in the figure, with the path shown as bold. According to Hamilton's principle of least action (2.8), the action integral S_{ba} is stationary with respect to first-order variations about this classical path. Consequently, all paths

2 in the immediate vicinity of path 1 have approximately the same phase for φ_{ba}. The waves for these paths therefore interfere constructively.

For paths which are remote from the classical path, a small variation in path leads to a large variation in phase. This is represented by paths 3 and 4 in Figure 3.2. The phase factor in the expression for φ_{ba} oscillates rapidly for these paths. The sum over these paths therefore is very close to zero on average. Only paths in the immediate vicinity of the classical path contribute significantly to the overall amplitude $K(x_b, t_b; x_a, t_a)$. Quantitatively, the action integrals S_{ba} for the two nearby paths 1 and 2 can at most differ by a reasonably small fraction of $\hbar$ for constructive interference to occur.

Further physical insight can be gained by noticing that, for any single path

$$\varphi_{ba} = \varphi_{bc} \cdot \varphi_{ca}, \tag{3.135}$$

where (x_c, t_c) is *any* intermediate space-time point along the particular path. This arises from (3.132), which leads directly to

$$S_{ba} = S_{bc} + S_{ca}. \tag{3.136}$$

This is depicted for a hypothetical system in Figure 3.3. The motion can be decomposed into a path from (x_a, t_a) to an intermediate point (x_c, t_c), followed by a path from (x_c, t_c) to the end point at (x_b, t_b). Since the kernel $K(x_b, t_b; x_a, t_a)$ is the integral over all possible paths, it follows that

$$K(x_b, t_b; x_a, t_a) = \int_{-\infty}^{\infty} K(x_b, t_b; x_c, t_c) \, K(x_c, t_c; x_a, t_a) \, dx_c. \tag{3.137}$$

This represents the motion between a specific starting point (x_a, t_a) and a specific end point (x_b, t_b).

For many purposes it is sufficient to know the state of the system at a given end point (x_b, t_b), without regard for the prior history of how the system got there. To this end we define the *wave function*

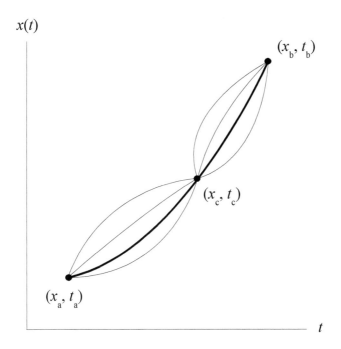

Figure 3.3: Evolution of possible paths through an intermediate point.

$\psi(x_b, t_b)$ as

$$\psi(x_b, t_b) = K(x_b, t_b; x_a, t_a), \qquad (3.138)$$

that is, we simply ignore the fact that the motion started at a particular point (x_a, t_a). Taking this assumption into account, it follows that

$$\psi(x_b, t_b) = \int_{-\infty}^{\infty} K(x_b, t_b; x_a, t_a)\, \psi(x_a, t_a)\, dx_a, \qquad (3.139)$$

where we have relabeled the indices.

Incidentally, this concept can be extended to any number of intermediate points, including a large number of points spaced infinitesimally close to one another. This leads to a method of more rigorously evaluating the path integral. The reader is referred to [29] for the mathematical details.

In words, (3.139) states that the wave function at any given point in space-time represents the summation over all possible prior histories. In addition, it follows from (3.134) that the probability density $P(x_b, t_b)$ for finding the particle at position x_b at time t_b is given by

$$P(x_b, t_b) = |\psi(x_b, t_b)|^2. \tag{3.140}$$

Given this, it is of great interest to explore the evolution of the wave function in a differential sense, where the end time t_b differs from the initial time t_a by a differential time interval ϵ. By this method we set out to derive a differential equation which describes the evolution of the wave function $\psi(x, t)$ in space-time. Applying (3.139) it follows that

$$\psi(x, t + \epsilon) = \int_{-\infty}^{\infty} K(x, t + \epsilon; x_a, t) \, \psi(x_a, t) \, dx_a. \tag{3.141}$$

Because this represents an infinitesimal increment in space-time, it follows that virtually all of the contribution is due to paths in the immediate vicinity of (x, t). We therefore make the substitution $x_a = x + \eta$, where η is a small increment in position relative to x. Substituting into (3.141) we obtain

$$\psi(x, t + \epsilon) = \int_{-\infty}^{\infty} K(x, t + \epsilon; x + \eta, t) \, \psi(x + \eta, t) \, d\eta. \tag{3.142}$$

The kernel K is given to good approximation by

$$K(x, t + \epsilon; x + \eta, t) = \frac{1}{A} \exp \left[\frac{i}{\hbar} \int_{t}^{t+\epsilon} L(x, v; t') \, dt' \right], \tag{3.143}$$

where A is a normalization constant, yet to be determined. For the infinitesimal integration interval this in turn reduces to

$$K(x, t + \epsilon, x + \eta, t) \approx \frac{1}{A} \exp \left[\frac{i}{\hbar} \epsilon L \left(x + \frac{\eta}{2}, \frac{\eta}{\epsilon} \right) \right], \tag{3.144}$$

where the first argument of L is the position and the second argument is the velocity, both averaged over the infinitesimal integration interval. The nonrelativistic approximation for the Lagrangian (2.7) is

$$L(x, v; t) = \tfrac{1}{2} m v^2 + q \mathbf{v} \cdot \mathbf{A}(x, t) - q\phi(x, t). \tag{3.145}$$

We assume for now that $\mathbf{A} = 0$ for the magnetic vector potential. Substituting, the wave function becomes

$$\psi(x, t + \epsilon) = \frac{1}{A} \int_{-\infty}^{\infty} \exp\left[\frac{im\eta^2}{2\hbar\epsilon}\right]$$
$$\cdot \exp\left[-\frac{i}{\hbar} \epsilon q \phi \left(x + \frac{\eta}{2}, t\right)\right] \psi(x + \eta, t) \, d\eta,$$

(3.146)

where most of the contribution to the integral is for small values of η. Next we expand $\psi(x, t)$ in a power series to first order in ϵ and second order in η. This gives

$$\psi(x, t) + \epsilon \frac{\partial \psi}{\partial t} = \frac{1}{A} \int_{-\infty}^{\infty} \exp\left[\frac{im\eta^2}{2\hbar\epsilon}\right] \cdot \left[1 - \frac{i}{\hbar} \epsilon V(x, t)\right]$$
$$\cdot \left[\psi(x, t) + \eta \frac{\partial \psi}{\partial x} + \frac{\eta^2}{2} \frac{\partial^2 \psi}{\partial x^2}\right] d\eta. \quad (3.147)$$

Equating the leading terms on both sides, we must have, to zero order in ϵ

$$\psi(x, t) = \psi(x, t) \cdot \frac{1}{A} \int_{-\infty}^{\infty} \exp\left[\frac{im\eta^2}{2\hbar\epsilon}\right] d\eta$$
$$= \psi(x, t) \cdot \frac{1}{A} \left(\frac{2\pi i\hbar\epsilon}{m}\right)^{1/2}. \quad (3.148)$$

Consequently, the normalization constant A is given by

$$A = \left(\frac{2\pi i\hbar\epsilon}{m}\right)^{1/2}. \quad (3.149)$$

Continuing to evaluate the right-hand side of (3.148), we make use of the two integrals

$$\frac{1}{A} \int_{-\infty}^{\infty} \eta \exp\left[\frac{im\eta^2}{2\hbar\epsilon}\right] d\eta = 0$$

$$\frac{1}{A} \int_{-\infty}^{\infty} \eta^2 \exp\left[\frac{im\eta^2}{2\hbar\epsilon}\right] d\eta = \frac{i\hbar\epsilon}{m}. \quad (3.150)$$

Substituting, we obtain

$$\psi(x,t) + \epsilon \frac{\partial}{\partial t} \psi(x,t) = \psi(x,t) - \frac{i}{\hbar} \epsilon q \phi(x,t) \psi(x,t) + \frac{i\hbar\epsilon}{2m} \frac{\partial^2}{\partial x^2} \psi(x,t).$$
(3.151)

Equivalently,

$$\left[-\frac{\hbar^2}{2m} \frac{\partial^2}{\partial x^2} + q\phi(x,t) \right] \psi(x,t) = i\hbar \frac{\partial}{\partial t} \psi(x,t).$$
(3.152)

We recognize this as the time-dependent Schrödinger equation (3.13) for one spatial dimension. Since the wave function $\psi(x,t)$ is itself a kernel, it follows that the kernel $K(x_b, t_b; x_a, t_a)$ satisfies Schrödinger's equation as well. It is straightforward to generalize the above arguments to three spatial dimensions, in which case one obtains the full time-dependent Schrödinger equation (3.13) in three spatial dimensions.

This shows the connection between the path integral approach and the more traditional approach. It also vindicates our initial choice for the form (3.133) of the amplitude φ_{ba}. For a charged particle optics system of macroscopic dimensions, only paths infinitesimally close to the classical path of motion, including the classical path itself, contribute significantly to the wave function.

It should be added in this context that the path integral approach is quite general, and applies to systems of atomic dimensions as well as systems of macroscopic dimensions. In an atomic system, the path integral is of the order of $\hbar$ for all paths. Consequently, all paths must be included in the path integral. This highlights the simplification which is possible for a charged particle system of macroscopic dimensions.

In most cases it is simpler to solve a differential equation than to perform the path integral. In the next section we investigate solutions to the Schrödinger equation for a single charged particle in a general electromagnetic potential.

Problems

1. Show that, for a free particle in one spatial dimension, the classical action integral S_{ba} given by (2.10) can be evaluated in closed form in the non-relativistic approximation as

$$S_{ba} = \frac{m}{2} \frac{(x_b - x_a)^2}{t_b - t_a}. \tag{3.153}$$

2. For a free particle, the kernel $K_0(x_b, t_b; x_a, t_a)$ can be evaluated in principle by subdividing the interval $(x_b, t_b; x_a, t_a)$ into N subintervals of equal time step ϵ. Summing over all possible paths, this leads to

$$K_0(x_b, t_b; x_a, t_a) = \lim_{\epsilon \to 0} \frac{1}{A} \int \int \cdots \int \exp\left(\frac{iS_{ba}}{\hbar}\right)$$
$$\cdot \frac{dx_1}{A} \frac{dx_2}{A} \cdots \frac{dx_{N-1}}{A}, \tag{3.154}$$

where A is given by (3.149), and S_{ba} is the free-particle action integral from the preceding problem. Show by repeated integration that the free-particle kernel K_0 can be expressed in closed form as

$$K_0(x_b, t_b; x_a, t_a) = \left[\frac{m}{2\pi i \hbar (t_b - t_a)}\right]^{1/2} \exp\left[\frac{im(x_b - x_a)^2}{2\hbar(t_b - t_a)}\right]. \tag{3.155}$$

Note that the probability density P_{ba} that the particle arrives at (x_b, t_b) is proportional to the absolute square of the kernel K_0. This is

$$P_{ba}(x_b, t_b; x_a, t_a) = \frac{m}{2\pi\hbar(t_b - t_a)}. \tag{3.156}$$

(Hint: the integral of a Gaussian function is also a Gaussian function. See [29, page 42] for detailed discussion.)

3. Show that, in three Cartesian dimensions, the wave function $\psi(\mathbf{x}_b, t_b)$ satisfies the three-dimensional time-dependent Schrödinger equation (3.13). (Hint: this is a straightforward generalization of the derivation for one dimension.)

3.2.2 Series solution for a particle in a general electromagnetic potential

The central problem in this section is to solve for the single-particle wave function $\psi(\mathbf{x}, t)$ in three spatial dimensions, in the presence of a general electromagnetic potential with components $\mathbf{A}(\mathbf{x}, t)$ and $\phi(\mathbf{x}, t)$. All relevant electromagnetic effects are contained in these potentials. All relevant quantum-mechanical information about the particle is contained in the wave function $\psi(\mathbf{x}, t)$.

Applying (3.4–3.13) we write the generalized time-dependent (non-relativistic) Schrödinger equation as

$$i\hbar \frac{\partial}{\partial t} \psi(\mathbf{x}, t) = \frac{1}{2m} \left[-i\hbar \nabla - q\, \mathbf{A}(\mathbf{x}, t) \right]^2 \psi(\mathbf{x}, t) + q\, \phi(\mathbf{x}, t)\, \psi(\mathbf{x}, t).$$

(3.157)

Applying the square bracket twice in succession, we obtain

$$\left[-i\hbar \nabla - q\, \mathbf{A}(\mathbf{x}, t) \right]^2 \psi(\mathbf{x}, t)$$
$$= \left[-\hbar^2\, \nabla^2 + 2i\hbar q \mathbf{A} \cdot \nabla + i\hbar q\, (\nabla \cdot \mathbf{A}) + q^2 \mathbf{A}^2 \right] \psi(\mathbf{x}, t).$$

(3.158)

This leads to the wave equation

$$i\hbar \frac{\partial}{\partial t} \psi(\mathbf{x}, t)$$
$$= \frac{1}{2m} \left[-\hbar^2\, \nabla^2 + 2i\hbar q \mathbf{A} \cdot \nabla + i\hbar q\, (\nabla \cdot \mathbf{A}) + q^2 \mathbf{A}^2 \right] \psi(\mathbf{x}, t)$$
$$+ q\, \phi\, \psi(\mathbf{x}, t),$$

(3.159)

which immediately reduces to the time-dependent Schrödinger equation (3.11) in the limit $\mathbf{A} = 0$ where no magnetic effects are present.

We seek a form for the wave function $\psi(\mathbf{x}, t)$ which approximates a free particle in the case where the electromagnetic potentials $\mathbf{A}(\mathbf{x}, t)$ and $\phi(\mathbf{x}, t)$ are slowly varying in space-time. To this end we *assume* that the solution $\psi(\mathbf{x}, t)$ can be expressed in the form

$$\psi(\mathbf{x}, t) = \exp\left[\frac{i}{\hbar} S(\mathbf{x}, t)\right], \qquad (3.160)$$

where $S(\mathbf{x}, t)$ has yet to be determined. There is no loss of generality, assuming $S(\mathbf{x}, t)$ is complex, and remembering that $\psi(\mathbf{x}, t)$ can be multiplied by an arbitrary normalization constant without affecting the validity of the solution. We can write down a few useful identities as follows:

$$\begin{aligned} \nabla \psi &= \frac{i}{\hbar} (\nabla S) \psi \\ \nabla^2 \psi &= \left[-\frac{1}{\hbar^2} (\nabla S)^2 + \frac{i}{\hbar} \nabla^2 S\right] \psi \\ \frac{\partial}{\partial t} \psi &= \left(\frac{i}{\hbar} \frac{\partial S}{\partial t}\right) \psi. \end{aligned} \qquad (3.161)$$

Substituting these into (3.135), it is straightforward to show that $S(\mathbf{x}, t)$ satisfies

$$(\nabla S - q\mathbf{A})^2 - i\hbar \nabla \cdot (\nabla S - q\mathbf{A}) + 2m \left(\frac{\partial S}{\partial t} + q\phi\right) = 0. \quad (3.162)$$

The second term on the left is obviously proportional to $\hbar$. For a single particle in an unbound state, we can regard this term as small relative to the other terms. (This will be justified later.) In this case we can approximate

$$(\nabla S - q\mathbf{A})^2 + 2m \left(\frac{\partial S}{\partial t} + q\phi\right) \approx 0. \qquad (3.163)$$

We immediately notice the striking fact that this is precisely the classical Hamiltonian–Jacobi equation of motion (2.313). We conclude from this that the function $S(\mathbf{x}, t)$ is approximately identified with Hamilton's principal function.

Equation (3.162) is nonlinear, and as such cannot be solved in closed form. We therefore seek a suitable approximation. To this end, we write $S(\mathbf{x}, t)$ as an infinite series,

$$S(\mathbf{x}, t) = S_0(\mathbf{x}, t) + \hbar S_1(\mathbf{x}, t) + \hbar^2 S_2(\mathbf{x}, t) + \dots. \qquad (3.164)$$

Substituting, and collecting terms in the various powers of $\hbar$, we obtain

$$
\begin{aligned}
0 = & \left[(\nabla S_0 - q\mathbf{A})^2 + 2m \left(\frac{\partial S_0}{\partial t} + q\phi \right) \right] \\
+ & \; \hbar \left[2\nabla S_1 \cdot (\nabla S_0 - q\mathbf{A}) - i\nabla \cdot (\nabla S_0 - q\mathbf{A}) + 2m \frac{\partial S_1}{\partial t} \right] \\
+ & \; \hbar^2 \left[2\nabla S_2 \cdot (\nabla S_0 - q\mathbf{A}) - i\nabla^2 S_1 + (\nabla S_1)^2 + 2m \frac{\partial S_2}{\partial t} \right] \\
+ & \; O\left(\hbar^3 \right).
\end{aligned}
\tag{3.165}
$$

In order for this series to converge to a sensible result, the individual terms must become successively smaller. Physically, we expect that the motion must approach the classical motion if we regard $\hbar$ to approach zero. Anticipating passage to the classical limit, we therefore regard $\hbar$ to be small, but variable. This requires that each of the quantities in square brackets must vanish separately, thus leading to a set of coupled equations for $S_0, S_1, S_2, \ldots$.

Taking the first equation in the series, we write

$$
(\nabla S_0 - q\mathbf{A})^2 + 2m \left(\frac{\partial S_0}{\partial t} + q\phi \right) = 0,
\tag{3.166}
$$

recalling that we regard the potentials $\phi(\mathbf{x}, t)$ and $\mathbf{A}(\mathbf{x}, t)$ to be functions of position $\mathbf{x}$ and time t.

Next we seek the solution for $S_0(\mathbf{x}, t)$. Rearranging terms, we can write this as

$$
\frac{\partial S_0}{\partial t} = -\frac{1}{2m} (\nabla S_0 - q\mathbf{A})^2 - q\phi.
\tag{3.167}
$$

The right-hand side is recognizable as the negative of the classical Hamiltonian H, where we make the identification

$$
\nabla S_0 = \mathbf{P},
\tag{3.168}
$$

and $\mathbf{P}$ is the canonical momentum. In this approximation, (3.166) is precisely the classical Hamiltonian-Jacobi equation of motion

(2.313). We conclude from this that the function $S_0(\mathbf{x}, t)$ is identified with Hamilton's principal function. Based on the expression (2.327) for Hamilton's principal function, we are prompted to propose a solution for S_0 as follows:

$$S_0(\mathbf{x}, t; \mathbf{x}_a, t_a) = \int_{t_a}^{t} L(\mathbf{x}, \mathbf{v}; t') \, dt', \qquad (3.169)$$

where the right-hand side is the action integral in Hamilton's principle of least action. Substituting this solution into (3.166) and making use of (3.168), it is straightforward to verify that this is indeed the correct solution. Furthermore, substituting S_0 into (3.160) we recover (3.133) from the path integral approach described earlier.

Forming the second equation for S_1 from (3.165) we have

$$2\nabla S_1 \cdot (\nabla S_0 - q\mathbf{A}) - i\nabla \cdot (\nabla S_0 - q\mathbf{A}) + 2m\frac{\partial S_1}{\partial t} = 0. \quad (3.170)$$

Making use of (3.168) this reduces to

$$2\nabla S_1 \cdot \mathbf{p} - i\nabla \cdot \mathbf{p} + 2m\frac{\partial S_1}{\partial t}, \qquad (3.171)$$

where $\mathbf{p}$ is the kinetic momentum given by $\mathbf{p} = \mathbf{P} - q\mathbf{A}$. Assuming the potentials $\mathbf{A}(\mathbf{x}, t)$ and $\phi(\mathbf{x}, t)$ are slowly varying, we can approximate this as

$$2p\frac{\partial S_1}{\partial s} - i\frac{\partial p}{\partial s} + 2m\frac{\partial S_1}{\partial t} = 0, \qquad (3.172)$$

where s is the coordinate along the path of motion, to which the kinetic momentum $\mathbf{p}$ is locally tangent. This reduces to

$$\left(\frac{\partial}{\partial s} + \frac{m}{p}\frac{\partial}{\partial t}\right) S_1 = \frac{i}{2}\frac{\partial}{\partial s}(\ln p). \qquad (3.173)$$

This equation can be solved in principle for S_1. We assume that the further terms in the series (3.165) for S become progressively smaller for an unbound particle. This will be discussed in more detail later. In principle, we substitute the terms $S_0, S_1, S_2, \ldots$ into (3.160) to form the wave function $\psi(\mathbf{x}, t)$.

We now turn our attention to the important special case where the potentials $\mathbf{A}$ and ϕ have no explicit time dependence. In this case the potentials can be written as $\mathbf{A}(\mathbf{x})$ and $\phi(\mathbf{x})$, respectively. The earlier analysis showed that the Hamiltonian has no explicit time dependence in this case, from which it follows that the total energy H is conserved. According to Hamilton–Jacobi theory, the function S_0 can be expressed as

$$S_0(\mathbf{x}, t) = W_0(\mathbf{x}) - H\,t, \tag{3.174}$$

where W_0 is Hamilton's characteristic function. Noting that $\nabla W_0 = \nabla S_0$, we obtain

$$(\nabla W_0 - q\mathbf{A})^2 = 2m\,(H - q\phi). \tag{3.175}$$

We recognize the right side as the square of the kinetic momentum $[p(\mathbf{x})]^2$, where

$$[p(\mathbf{x})]^2 = 2m\,[\,H - q\phi(\mathbf{x})\,]. \tag{3.176}$$

This is satisfied by

$$\nabla W_0 - q\,\mathbf{A} = \pm\,\mathbf{p}(\mathbf{x}), \tag{3.177}$$

Retaining only the positive (right-propagating) root, and ignoring the negative (left-propagating) root, we obtain

$$\nabla W_0 = \mathbf{P}(\mathbf{x}), \tag{3.178}$$

recalling that $\mathbf{P} = \mathbf{p} + q\,\mathbf{A}$ is the canonical momentum. Integrating, we obtain

$$W_0(\mathbf{x}_b, t_b) - W_0(\mathbf{x}_a, t_a) = \int_{\mathbf{x}_a}^{\mathbf{x}_b} \mathbf{P} \cdot d\mathbf{x}, \tag{3.179}$$

where the integral is a line integral along a path joining the points $\mathbf{x}_a$ and $\mathbf{x}_b$. Substituting, this leads to

$$S_0(\mathbf{x}_b, t_b) = S_0(\mathbf{x}_a, t_a) + \int_{\mathbf{x}_a}^{\mathbf{x}_b} \mathbf{P} \cdot d\mathbf{x} - H(t_b - t_a). \qquad (3.180)$$

The second term in (3.165) leads to

$$2(\nabla S_0 - q\mathbf{A}) \cdot \nabla S_1 - i\nabla \cdot (\nabla S_0 - q\mathbf{A}) = 0. \qquad (3.181)$$

Substituting the solution for $S_0(\mathbf{x}, t)$, this becomes

$$\mathbf{p} \cdot \nabla S_1 = \frac{i}{2} \nabla \cdot \mathbf{p}. \qquad (3.182)$$

We now assume that the potentials $\mathbf{A}(\mathbf{x})$ and $\phi(\mathbf{x})$ do not vary significantly over distances comparable with the deBroglie wavelength $\lambda = h/p$. This allows the approximation

$$p \frac{\partial}{\partial s} S_1 = \frac{i}{2} \frac{\partial}{\partial s} p, \qquad (3.183)$$

where s represents the coordinate along the path of motion, to which the kinetic momentum vector $\mathbf{p}$ is locally tangent. This is equivalent to

$$\frac{\partial}{\partial s} S_1 = \frac{i}{2} \frac{\partial}{\partial s} (\ln p). \qquad (3.184)$$

We immediately perform the line integral to obtain

$$S_1(\mathbf{x}_b) - S_1(\mathbf{x}_a) = \frac{i}{2} [\ln p(\mathbf{x}_b) - \ln p(\mathbf{x}_a)] = i \ln \left[\frac{p(\mathbf{x}_b)}{p(\mathbf{x}_a)} \right]^{1/2}. \qquad (3.185)$$

Recalling the definition (3.164) for S, and substituting the results for S_0 and S_1, we obtain the solution for the wave function for time-independent potentials $\phi(\mathbf{x})$ and $\mathbf{A}(\mathbf{x})$ as

$$\psi(\mathbf{x}_b, t_b) = \psi(\mathbf{x}_a, t_a) \left[\frac{p(\mathbf{x}_a)}{p(\mathbf{x}_b)} \right]^{1/2} \exp \left[\frac{i}{\hbar} \int_{\mathbf{x}_a}^{\mathbf{x}_b} \mathbf{P} \cdot d\mathbf{x} - \frac{iH}{\hbar} (t_b - t_a) \right] \qquad (3.186)$$

recalling that H is the conserved total energy. We have ignored terms of order $\hbar^2$ in the expansion for S, since these are expected to be relatively small for a single particle in an unbound state.

This solution is known as the *WKB approximation* [59, 79]. It has an important, and still very relevant history in quantum mechanics, in explaining the classical limit of large quantum numbers. For a bound system the approximation breaks down at the classical turning points where the kinetic momentum $p(\mathbf{x}) = \sqrt{2m(H - q\phi)} = 0$. Physically, the WKB approximation applies in the case where the fractional change in the electrostatic potential $\phi(\mathbf{x})$ is small over a distance comparable with the deBroglie wavelength. For an unbound system at relatively high energy, the approximation is excellent. In the free-particle case where the potentials are zero everywhere, this solution reverts to the familiar plane wave solution as required.

It is important to remember that this solution applies to a multiplicity of paths, of which the classical trajectory is just one. These must be summed to obtain the overall wave function. This involves a procedure similar to (3.139), adapted to three dimensions. In most charged particle optical systems, the action integrals in the solutions (3.169, 3.186) are very large relative to $\hbar$. We showed in the preceding section that only trajectories infinitesimally separated from the classical trajectory, together with the classical trajectory itself, contribute appreciably to the overall wave function. In this case it is a very good approximation to assume that the action integrals are applied *only* along the classical trajectory.

This can be further understood by applying the operator for the canonical momentum $\mathbf{P}$ to the wave function (3.186). This gives

$$-i\hbar\nabla\,\psi(\mathbf{x}, t) = \mathbf{P}\,\psi(\mathbf{x}, t). \tag{3.187}$$

Geometrically, this means that the canonical momentum vector $\mathbf{P}$ is perpendicular to the surfaces of constant phase. The kinetic momentum vector $\mathbf{p}$ is everywhere tangent to the classical trajectory. In the presence of a magnetic vector potential $\mathbf{A}$, this gives rise to a geometrical interpretation as shown in Figure 3.4.

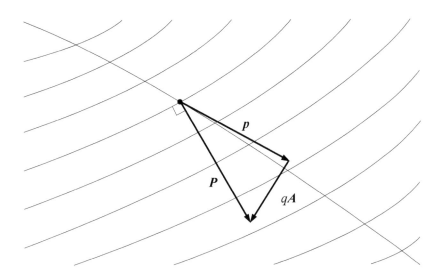

Figure 3.4: Particle momentum and the surfaces of constant phase.

The probability current $j(\mathbf{x}, t)$ is given by

$$j(\mathbf{x}) = \frac{i\hbar}{2m} (\psi \nabla \bar{\psi} - \bar{\psi} \nabla \psi), \qquad (3.188)$$

where the time drops out, as required in the case where the potentials have no explicit time dependence. Substituting (3.187) the reader can immediately deduce that

$$j(\mathbf{x}, t) = \frac{\mathbf{P}}{m} \bar{\psi} \psi. \qquad (3.189)$$

This is equivalent to the continuity equation of probability, for which the current density equals the velocity (momentum divided by mass) times the probability density. In the WKB approximation, we see immediately that

$$\mathbf{p}(\mathbf{x}) \bar{\psi} \psi = const. \qquad (3.190)$$

This interpretation suggests a practical approach to computing the wave function ψ. First, we compute the classical trajectory, given a prespecified initial position $\mathbf{x}$, velocity $\mathbf{v}$. Next, we compute the classical action integral between the point $(\mathbf{x}_a)$ and any other point $(\mathbf{x}_b)$. Finally, we divide by $\hbar$ to give the phase, thus obtaining the wave function (3.186). Incidentally, this approach is accurate in the sense that it implicitly includes all orders af aberrations.

3.2.3   Quantum interference effects in electromagnetic potentials

In the preceding sections we investigated wave-optical interference which occurs when single free-particle amplitudes corresponding to alternative paths of motion add coherently. We now extend this discussion to the case where electromagnetic potentials are present. We consider a hypothetical monochromatic point source of electrons at axial coordinate z_a, which coincides with the front focal plane of a lens at axial coordinate z_{L1}. This is shown schematically in Figure 3.5. A screen with two slits is located directly behind the lens. The slits are illuminated by a monochromatic plane wave in the paraxial approximation. A second lens at axial coordinate z_{L2} produces a diffraction pattern on a viewing screen located at axial coordinate z_b, which is assumed to coincide with the back focal plane of the second lens. The solid lines correspond to the classical rays for the two alternative paths. Bright fringes appear where the amplitudes corresponding to two alternative paths add constructively. This occurs where the optical path lengths differ by an integral number of wavelengths.

Next we assume a magnetic flux which is entirely confined to the cross-hatched circle, where the lines of flux are oriented perpendicular to the plane of the figure. Such a flux can be produced in principle by a very long solenoid with very fine, closely spaced windings, where the axis of the solenoid is also oriented perpendicular to the plane of the figure. We assume that the magnetic field

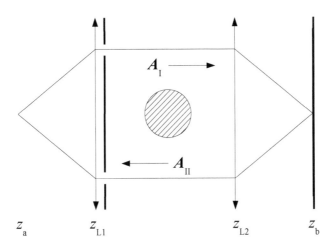

Figure 3.5: Two-slit interference in presence of a magnetic vector potential.

is zero outside the cross-hatched region, and that the flux region lies entirely within the geometric shadow of the two slits. It follows that the electron experiences no magnetic Lorentz force, since the magnetic field is zero wherever significant likelihood of finding the electron exists. Strictly speaking, these assumptions can only be approximately realized, since the flux lines must follow a return path outside the solenoid. The magnetic field can be made arbitrarily small by judicious design of the experimental configuration, however.

The amplitude $\psi(\mathbf{x}_b, t_b)$ is given in terms of the initial amplitude $\psi(\mathbf{x}_a, t_a)$ by (3.186)

$$\psi(\mathbf{x}_b, t_b) = \psi(\mathbf{x}_a, t_a) \left[\frac{p(\mathbf{x}_a)}{p(\mathbf{x}_b)} \right]^{1/2}$$
$$\cdot \exp\left\{ \frac{i}{\hbar} \left[\int_{\mathbf{x}_a}^{\mathbf{x}_b} \mathbf{P} \cdot d\mathbf{s} - H(t_b - t_a) \right] \right\} (3.191)$$

for the special case where the potentials $\mathbf{A}(\mathbf{x})$ and $\phi(\mathbf{x})$ have no explicit time dependence. We assume in the following that $p(\mathbf{x}_a) = p(\mathbf{x}_b)$; i.e., the kinetic momentum is the same at the start and end points. This is equivalent to the assumption that $\phi(\mathbf{x}_a) = \phi(\mathbf{x}_b)$; i.e., the electrostatic potential is the same at the start and end points. The resultant amplitude $\psi(\mathbf{x}_b, t_b)$ is the sum of the amplitudes for the two paths, namely,

$$\psi(\mathbf{x}_b, t_b) = \psi_I(\mathbf{x}_b, t_b) + \psi_{II}(\mathbf{x}_b, t_b), \tag{3.192}$$

where $\psi_I(\mathbf{x}_b, t_b)$ and $\psi_{II}(\mathbf{x}_b, t_b)$ are the amplitudes corresponding to the upper and lower paths in Figure 3.5, respectively. We can write this equivalently as

$$\psi(\mathbf{x}_b, t_b) = \psi(\mathbf{x}_a, t_a) \left(e^{i\theta_I} + e^{i\theta_{II}} \right), \tag{3.193}$$

where we have defined the phases

$$\begin{aligned} \theta_I &= \frac{1}{\hbar} \int_{\mathbf{x}_a}^{\mathbf{x}_b} \mathbf{P}_I \cdot d\mathbf{s}_I - \frac{1}{\hbar} H (t_b - t_a), \\ \theta_{II} &= \frac{1}{\hbar} \int_{\mathbf{x}_a}^{\mathbf{x}_b} \mathbf{P}_{II} \cdot d\mathbf{s}_{II} - \frac{1}{\hbar} H (t_b - t_a). \end{aligned} \tag{3.194}$$

The intensity in the plane of the screen z_b is given by

$$I(\mathbf{x}_b) = |\psi(\mathbf{x}_b, t_b)|^2. \tag{3.195}$$

It is straightforward to show that this is equivalent to

$$I(\mathbf{x}_b) = 4\, I(\mathbf{x}_a) \cos^2 \left(\frac{\theta_{II} - \theta_I}{2} \right). \tag{3.196}$$

The time dependence in (3.194) subtracts to zero, corresponding to a standing wave. Constructive interference occurs for $\theta_{II} - \theta_I = 2n\pi$, and destructive interference occurs for $\theta_{II} - \theta_I = (2n+1)\pi$, where the integer n represents the order. The phase difference is given by

$$\begin{aligned} \theta_{II} - \theta_I &= \frac{1}{\hbar} \oint \mathbf{P} \cdot d\mathbf{s} \\ &= \frac{1}{\hbar} \oint \mathbf{p} \cdot d\mathbf{s} + \frac{q}{\hbar} \oint \mathbf{A} \cdot d\mathbf{s}, \end{aligned} \tag{3.197}$$

where the integral is around the closed path. Next we define a phase shift $\Delta\theta$ corresponding to the difference in phase between the solenoid being excited to a specific value and the solenoid current turned off. This is

$$\Delta\theta = \frac{q}{\hbar} \oint \mathbf{A} \cdot d\mathbf{s}. \tag{3.198}$$

Applying Stokes's theorem we write

$$\begin{aligned} \Delta\theta &= \frac{q}{\hbar} \int_S (\nabla \times \mathbf{A}) \cdot d\mathbf{S} \\ &= \int_S \mathbf{B} \cdot d\mathbf{S}, \end{aligned} \tag{3.199}$$

where S is any surface bounded by the closed ray paths. Equivalently,

$$\Delta\theta = \frac{q\,\Phi}{\hbar}, \tag{3.200}$$

where Φ is the total magnetic flux enclosed by the ray paths. The magnetic vector potential is nonzero in the vicinity of the classical trajectories, but the magnetic field is zero there. Therefore, no magnetic Lorentz force acts on the electron. This result was first predicted by Ehrenberg and Siday [25], and later expanded upon by Aharonov and Bohm [2].

An electrostatic analog was first predicted by Aharonov and Bohm [2]. This is shown schematically in Figure 3.6. An electron traversing the upper path passes through a conducting tube. When the electron is inside the tube, near its center, an electrostatic potential $V(t)$ is momentarily applied to the tube by an external source. Assuming the length of the tube is much larger than its diameter, the electron experiences no electric field during the time interval in which $V(t)$ is switched on. Consequently, no electrostatic Lorentz force is exerted on the electron.

This is only possible in principle if the electron is represented by a wave packet, rather than a monochromatic plane wave. The wave packet must be sufficiently localized for the above condition to be met, whereas a monochromatic plane wave has infinite extent. This inevitably requires a spread in momentum as well, consistent with the Heisenberg uncertainty principle. Measurable intensity at

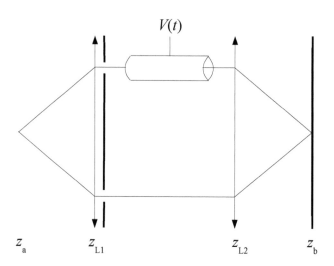

Figure 3.6: Two-slit interference in presence of an electrostatic potential.

z_b only occurs when the path difference is less than the coherence length of the wave packet.

Since the electrostatic potential $\phi(\mathbf{x}, t)$ depends explicitly on time, we must use the time-dependent formulation. The wave function $\psi(\mathbf{x}_b, t_b)$ is given (3.132, 3.145) by

$$\psi(\mathbf{x}_b, t_b) = \psi(\mathbf{x}_a, t_a) \, \exp\left[\frac{i}{\hbar} \int_{t_a}^{t_b} L(\mathbf{x}, \mathbf{v}; t)\, dt\right], \qquad (3.201)$$

where $L(\mathbf{x}, \mathbf{v}; t)$ is the classical Lagrangian given by (2.9) as

$$L(\mathbf{x}, \mathbf{v}; t) = -m\, c^2\, \sqrt{1 - v^2/c^2} + q\,\mathbf{v} \cdot \mathbf{A}(\mathbf{x}, t) - q\,\phi(\mathbf{x}, t). \quad (3.202)$$

The magnetic vector potential $\mathbf{A}$ is assumed to be zero in this case. The amplitude $\psi(\mathbf{x}_b, t_b)$ is again the sum of the amplitudes for the two alternative paths (3.192, 3.193), with the respective phases given by

$$\theta_I = \frac{1}{\hbar} \int_{t_a}^{t_b} L_I \, dt_I$$

$$\theta_{II} = \frac{1}{\hbar} \int_{t_a}^{t_b} L_{II} \, dt_{II}. \qquad (3.203)$$

The intensity measured in the plane at z_b again depends on the difference between these two phases. The phase shift between the cases with the potential $V(t)$ switched on and off is

$$\Delta\theta = \frac{q}{\hbar} \int V(t)\, dt. \qquad (3.204)$$

This is independent of the electron energy, and is therefore the same for all constituent energies in the wave packet.

The phase shifts (3.198, 3.204) result in a measurable lateral shift in the fringe pattern on the screen at z_b in principle. The solutions (3.191, 3.201) for the wave function have the striking property that they depend only on the magnetic vector potential $\mathbf{A}(\mathbf{x}, t)$, and the electrostatic scalar potential $\phi(\mathbf{x}, t)$. Nowhere do the magnetic field $\mathbf{B}$ or the electric field $\mathbf{E}$ appear. This is distinctly different from the classical description, in which these fields appear explicitly in the Lorentz force law (2.15). Indeed no Lorentz force is present in this quantum mechanical description. The reader is referred to [3, 87] for further elaboration, including a description of experimental results.

3.2.4   The Klein–Gordon equation and the covariant wave function

The effects of special relativity become important when the kinetic energy of the particle is comparable to, or greater than mc^2, where m is the rest mass. A correct treatment must also include the effects of spin. The reader is referred to the book by Bjorken and Drell [6].

In many practical instruments, spin does not play an important role in the optics, however. A useful approximation is available in the Klein–Gordon equation which ignores spin, but retains Lorentz covariance. As before, we confine our attention to the lab frame only.

Following the arguments of the preceding sections, we *assume* that

$$\psi(\mathbf{x}, t) = \psi(\mathbf{x}_a, t_a) \exp\left[\frac{i}{\hbar}\int_{t_a}^{t} L(\mathbf{x}, \mathbf{v}; t')\, dt'\right], \qquad (3.205)$$

where $L(\mathbf{x}, \mathbf{v}; t)$ is the relativistic classical Lagrangian given in the lab frame by (2.9) as

$$L(\mathbf{x}, \mathbf{v}; t) = -m\,c^2\,\sqrt{1 - v^2/c^2} + q\,\mathbf{v}\cdot\mathbf{A}(\mathbf{x}, t) - q\,\phi(\mathbf{x}, t). \quad (3.206)$$

From (2.25, 2.30, 2.31) the classical energy-momentum relationship is given by

$$[\,H - q\,\phi(\mathbf{x}, t)\,]^2 = [\,\mathbf{P} - q\,\mathbf{A}(\mathbf{x}, t)\,]^2 c^2 + m^2 c^4. \qquad (3.207)$$

This equation is Lorentz-invariant, since it contains the square of the difference of two four vectors $(\mathbf{P}, iH/c)$ and $(q\mathbf{A}, iq\phi/c)$. As such, it has the same form in every uniformly moving reference frame. Again invoking the fundamental postulate that classical quantities are replaced by their quantum mechanical operators, this becomes

$$\left(i\hbar\frac{\partial}{\partial t} - q\,\phi\right)^2 \psi(\mathbf{x}, t) = (-i\hbar\nabla - q\mathbf{A})^2\,c^2\,\psi(\mathbf{x}, t) + m^2 c^4 \psi(\mathbf{x}, t).$$

$$(3.208)$$

Applying the operator in large parentheses twice in succession, the left side is

$$\left(i\hbar\frac{\partial}{\partial t} - q\,\phi\right)^2 \psi(\mathbf{x}, t)$$

$$= -\hbar^2\frac{\partial^2\psi}{\partial t^2} - 2i\hbar q\,\phi\,\frac{\partial\psi}{\partial t} - i\hbar q\,\frac{\partial\phi}{\partial t}\,\psi + q^2\phi^2\psi. \quad (3.209)$$

Similarly, the first term on the right is

$$(-i\hbar\nabla - q\mathbf{A})^2\,c^2\,\psi(\mathbf{x}, t)$$

$$= -\hbar^2 c^2\,\nabla^2\psi + 2i\hbar q c^2\,\mathbf{A}\cdot\nabla\psi + i\hbar q c^2\,(\nabla\cdot\mathbf{A})\,\psi + q^2 c^2\,\mathbf{A}^2\,\psi.$$

$$(3.210)$$

Substituting, we obtain the relativistic time-dependent wave equation as follows:

$$-\hbar^2 \frac{\partial^2 \psi}{\partial t^2} - 2i\hbar q\,\phi\,\frac{\partial \psi}{\partial t} - i\hbar q\,\frac{\partial \phi}{\partial t}\,\psi + q^2\phi^2\psi$$

$$= -\hbar^2 c^2\,\nabla^2\psi + 2i\hbar qc^2\,\mathbf{A}\cdot\nabla\psi + i\hbar qc^2\,(\nabla\cdot\mathbf{A})\,\psi$$

$$+q^2c^2\,\mathbf{A}^2\,\psi + m^2c^4.$$

$$(3.211)$$

Grouping terms and dividing through by c^2, we obtain

$$-\hbar^2\left(\nabla^2\psi - \frac{1}{c^2}\frac{\partial^2\psi}{\partial t^2}\right) + 2i\hbar q\left(\mathbf{A}\cdot\nabla\psi + \frac{\phi}{c^2}\frac{\partial\psi}{\partial t}\right)$$

$$+ i\hbar q\left(\nabla\cdot\mathbf{A} + \frac{1}{c^2}\frac{\partial\phi}{\partial t}\right)\psi + q^2\left(\mathbf{A}^2 - \frac{1}{c^2}\phi^2\right)\psi$$

$$+ m^2c^2\psi = 0. \qquad (3.212)$$

Again, we assume that $\psi(\mathbf{x}, t)$ can be written as

$$\psi(\mathbf{x}, t) = \exp\left[\frac{i}{\hbar}\,S(\mathbf{x}, t)\right]. \qquad (3.213)$$

Substituting this into the relativistic wave equation, it is tedious but straightforward to show that

$$(\nabla S - q\,\mathbf{A})^2 - i\hbar\nabla\cdot(\nabla S - q\,\mathbf{A})$$

$$-\frac{1}{c^2}\left(\frac{\partial S}{\partial t} + q\,\phi\right)^2 + \frac{i\hbar}{c^2}\frac{\partial}{\partial t}\left(\frac{\partial S}{\partial t} + q\,\phi\right) + m^2c^2 = 0.$$

$$(3.214)$$

Again we expand $S(\mathbf{x}, t)$ in powers of $\hbar$, in which we define

$$S(\mathbf{x}, t) = S_0(\mathbf{x}, t) + \hbar\,S_1(\mathbf{x}, t) + \hbar^2\,S_2(\mathbf{x}, t) + \ldots. \qquad (3.215)$$

Substituting and grouping terms according to powers of $\hbar$, this leads after some algebra to

$$
\begin{aligned}
0 \;=\; & \left[(\nabla S_0 - q\mathbf{A})^2 - \frac{1}{c^2} \left(\frac{\partial S_0}{\partial t} + q\,\phi \right)^2 + m^2 c^2 \right] \\
& + \;\hbar \;\left[(2\nabla S_1 - i\,\nabla) \cdot (\nabla S_0 - q\mathbf{A}) \right] \\
& - \;\hbar \left[\frac{1}{c^2} \left(2\frac{\partial S_1}{\partial t} + i\frac{\partial}{\partial t} \right) \left(\frac{\partial S_0}{\partial t} + q\,\phi \right) \right] \\
& + \; O\!\left(\hbar^2 \right).
\end{aligned}
\tag{3.216}
$$

Again anticipating the classical limit where $\hbar \to 0$, we set each of the coefficients of the powers of $\hbar$ equal to zero. This leads to the coupled set of equations as before. Taking the first equation in the series, we have

$$
(\nabla S_0 - q\mathbf{A})^2 - \frac{1}{c^2} \left(\frac{\partial S_0}{\partial t} + q\,\phi \right)^2 + m^2 c^2 = 0.
\tag{3.217}
$$

We now consider the special case that the potentials are time-independent. Again defining a new function $W_0(\mathbf{x})$ given by

$$
S_0(\mathbf{x}, t) = W_0(\mathbf{x}) - H\,t,
\tag{3.218}
$$

where H is the constant, conserved total energy eigenvalue. Substituting, this gives

$$
(\nabla W_0 - q\mathbf{A})^2 = \frac{1}{c^2} \left(-H + q\,\phi \right)^2 - m^2 c^2,
\tag{3.219}
$$

where we note that $\nabla W_0 = \nabla S_0$. We now make use of the relativistic energy-momentum relation

$$
(H - q\,\phi)^2 = p^2 c^2 + m^2 c^4
\tag{3.220}
$$

to obtain

$$
(\nabla W_0 - q\,\mathbf{A})^2 = [\,p(\mathbf{x})\,]^2,
\tag{3.221}
$$

where p is now the relativistic scalar kinetic momentum. This is satisfied by

$$
\nabla W_0 = \mathbf{P}(\mathbf{x}),
\tag{3.222}
$$

which we immediately recognize from the relationship (2.25) between the canonical momentum $\mathbf{P}$ and the kinetic momentum $\mathbf{p}$. We have neglected the negative root. This means that we only consider motion in the forward direction for the present purpose. Integrating, we obtain

$$S_0(\mathbf{x}_b, t_b) = S_0(\mathbf{x}_a, t_a) + \int_{\mathbf{x}_a}^{\mathbf{x}_b} \mathbf{P} \cdot d\mathbf{x} - H(t_b - t_a). \qquad (3.223)$$

We immediately recognize this set as being identical with the non-relativistic approximation, except that the relativistic quantities $\mathbf{p}, \mathbf{P}$, and H here replace their non-relativistic counterparts used previously. Noting that

$$\frac{\partial S_1}{\partial t} = \frac{\partial S_2}{\partial t} = \ldots = 0, \qquad (3.224)$$

the preceding analysis applies, and we obtain the solution for the wave function $\psi(\mathbf{x}, t)$ for the case of time-independent potentials as

$$\psi(\mathbf{x}_b, t_b) = \psi(\mathbf{x}_a, t_a) \left[\frac{p(\mathbf{x}_a)}{p(\mathbf{x}_b)} \right]^{1/2}$$
$$\cdot \exp\left\{ \frac{i}{\hbar} \left[\int_{\mathbf{x}_a}^{\mathbf{x}_b} \mathbf{P} \cdot d\mathbf{x} - H(t_b - t_a) \right] \right\}. \qquad (3.225)$$

Given an initial condition $\psi(\mathbf{x}_a, t_a)$ this describes the propagation of $\psi(\mathbf{x}_b, t_b)$ to any end point in the presence of static fields, where H is the conserved total energy. This represents a single eigenstate corresponding to the energy H. As before, individual eigenstates are linearly superimposed to build up the state function $\Psi(\mathbf{x}, t)$, which reflects the experimental preparation of the beam. The measurable intensity is given by $|\Psi(\mathbf{x}, t)|^2$.

In the free-particle case where $\phi = 0$ and $\mathbf{A} = 0$, the equation (3.212) reduces to

$$\left(\nabla^2 - \frac{1}{c^2} \frac{\partial^2}{\partial t^2} - \frac{m^2 c^2}{\hbar^2} \right) \psi(\mathbf{x}, t) = 0. \qquad (3.226)$$

This is known as the Klein–Gordon equation. This has non-normalized plane wave solutions given by

$$\psi(\mathbf{x}, t) = \exp\left[\frac{i}{\hbar}\left(\pm\mathbf{p}\cdot\mathbf{x} - Ht\right)\right]. \qquad (3.227)$$

The reader can verify that this is the correct solution direct substitution into the Klein–Gordon equation. One can also verify that $H^2 = p^2c^2 + m^2c^4$ for the free-particle case with $\phi = 0$ and $\mathbf{A} = 0$. The interpretation of $|\psi(\mathbf{x}, t)|^2$ in terms of probability density is more subtle than in the nonrelativistic approximation. At modest energies, $|\psi|^2$ remains a very good approximation to the relativistic probability density, however. The reader is referred to Bjorken and Drell [6] for a detailed discussion.

In summary, all relevant information about quantum mechanical particle motion in general, time-independent potentials $\mathbf{A}(\mathbf{x})$ and $\phi(\mathbf{x})$ is contained in the relativistic wave function (3.225).

3.2.5   Physical interpretation of the wave function and its practical application

The central problem in optics is to understand the intensity distribution in a given transverse plane of an optical system. This might be the image plane of an electron microscope, the Fourier plane of a diffractometer, or the dispersion plane of an energy-dispersive charged particle spectrometer, to name a few examples. The intensity distribution is proportional to the probability distribution for finding a single particle at a given position. This in turn is given by the absolute square of the wave function, which we have called $\psi(\mathbf{x}, t)$. In the preceding analysis we have focused on calculating the wave function for a single charged particle moving in a general electromagnetic potential. The aim of the present section is to understand how to translate this into an areal intensity distribution, such as one would measure in a practical instrument.

In classical geometrical optics, the beam can be regarded as a family of closely spaced trajectories. Each individual trajectory is calculated by solving the Euler–Lagrange equations of motion, which are in turn derived from Hamilton's principle of least action. The collective properties of these trajectories immediately lead to conservation of phase space volume, which in turn leads to the law of Helmholtz–Lagrange and brightness conservation, as derived in Chapter 2.

In quantum mechanical wave optics, one starts with a surface of constant phase called a *wave front*. This forms an initial condition, from which the wave propagates through space-time. This is shown schematically in Figure 3.7. At time t the wave front is

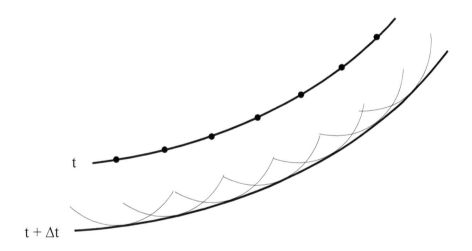

Figure 3.7: Huygens' principle.

depicted by the upper curve. At a later time $t + \Delta t$ the wave front has propagated, forming a new curve. The wave fronts are actually surfaces in three-dimensional coordinate space. In the figure we depict a planar slice through the wave front, which is a curve on the page.

We imagine a collection of point sources distributed over the initial wave front, with each point source radiating a spherical wave. The point sources are assumed to be infinite in number, and infinitesimally separated along the initial wave front. They are also assumed to radiate in phase, or *coherently* relative to one another. At the time $t + \Delta t$ the spherical waves have propagated to form the envelope of the new wave front. This equivalent picture of wave propagation is known as *Huygens' principle*. It will prove to be indispensible to formulating a mathematical description of diffraction, which is derived in later sections.

Next, we consider each point source to form the initial point of a wave function $\psi(\mathbf{x}_a, t_a)$. From the preceding analysis, the final state wave function $\psi(\mathbf{x}_b, t_b)$ is calculated from the solutions (3.169, 3.186), depending on whether the electromagnetic potentials have explicit time dependence or not. The wave functions corresponding to the separate point sources add coherently to form the composite wave front. The fact that the composite wave front has a specific curvature says that the point sources have a corresponding relative position in space-time, as well as a specific phase relationship to one another. This phase advances monotonically through space-time, as given by the action integral divided by $\hbar$. Each point source has an associated classical trajectory, as depicted schemaically in Figure 3.4.

The mathematical description is exact in principle, and accounts for all aberrations. In geometrical optics, the aberrations are manifest as a displacement of the classical particle trajectory from the paraxial approximation. This displacement can be calculated in principle to an arbitrary degree of accuracy. In wave optics, the aberrations are manifest as displacements in the surfaces of constant phase.

The intensity is proportional to the probability density, which in turn is related to the wave function as $|\psi(\mathbf{x}, t)|^2$. Obviously the phase does not appear explicitly here. It is the relative phases of neighboring trajectories that govern the shape of the wave fronts. As a probability, the wave function for each point source must satisfy

$$\int |\psi(\mathbf{x}, t)|^2 \, d^3\mathbf{x} = 1. \tag{3.228}$$

This implies a normalization constant multiplying the wave function. As Feynman and Hibbs point out [29], there seems to be no simple general procedure for calculating this constant. Even for the simple case of a free-particle plane wave, we had to resort to the device of periodic boundary conditions. Fortunately, the absolute probability is unimportant here. What is important is the *relative* probability for the various point sources. This determines the relative intensity across the beam. This becomes part of specifying the initial condition for each point source.

The fact that each point source radiates a spherical wave is equivalent to the initial momentum direction being completely unspecified. From the Heisenberg uncertainty principle, this is consistent with the initial point $(\mathbf{x}_a, t_a)$ being precisely specified for each point source. The initial *longitudinal* momentum is precisely specified for each classical trajectory. However, the Heisenberg principle has no classical analog. We must also remember that both the classical trajectory and the quantum mechanical wave function both apply to a single particle.

This completes the physical picture which connects the wave function to the intensity distribution of a practical system. We are now in a position to discuss the intensity distribution for a given practical system in a more general way, through the theory of diffraction. This forms the topic of the following sections.

3.3 Diffraction

Diffraction is the phenomenon which results from the propagation, spreading, and interference of waves. In experimental optics, interference is manifest as alternating bright and dark intensity bands called *fringes*. This is a purely wave-optical phenomenon with no analog in classical mechanics. Its origin dates back several hundred years to the pioneering work of da Vinci, Grimaldi, and Huygens.

In modern terms, the motion of a single charged particle in an electromagnetic potential is described quantum mechanically by a wave function, for which the absolute square is the probability density that a single measurement will find the particle at precise space-time coordinates. A state which is a superposition of two or more eigenstates with identical energy, but differing directions of momentum exhibits interference. Diffraction and interference are fundamental to a complete description of charged particle optics.

The purpose of this section is to place the concept of diffraction on a firm conceptual and mathematical basis, and then, based on this, to describe several useful examples. Before embarking on this, it is worthwhile to convey an intuitive feel for the subject by considering a simple thought experiment, which was described by Feynman, et. al. [30, Chapter 1, Volume 3]. This is shown schematically in Figure 3.8. We imagine a single charged particle with precisely known momentum and energy, incident perpendicularly on an opaque screen S with two parallel slits. We assume that the transverse position of the particle is completely unknown. Consequently, the particle could be stopped by the screen, or it could pass through one of the two slits. Assuming it passes through one of the slits, it is impossible to know which slit the particle passed through. Having passed through one of the slits, the particle drifts to a phosphor screen P at the bottom of the figure, where a flash of light is emitted on impact. This represents a measurement of the transverse position of the single particle on the phosphor screen. By itself, this measurement does not reveal much information, since the particle could land practically anywhere. This is corroborated by the fact that a second particle generally lands at a different place from the first particle.

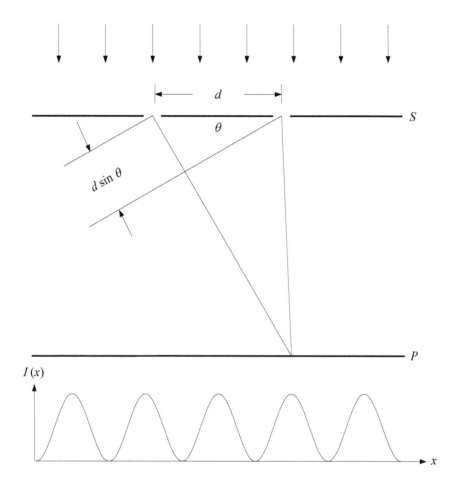

Figure 3.8: Two-slit thought experiment.

Next, we repeat this measurement for many, many individual particles. This is easily accomplished by forming a beam of particles.
For sufficiently low beam current, the individual beam particles
are far enough apart on average that they do not interact with one
another. One by one, particles arrive at the screen and produce
a flash of light. This is conceptually equivalent to repeating the
measurement many times, once for each particle, with identical

preparation of the quantum mechanical state for each measurement. The remarkable result is that bright and dark fringes are observed on the phosphor screen. This is represented by the intensity distribution as a function of transverse position x plotted at the bottom of the figure. This result has been observed directly for a variety of particle species, indicating that this is more than just a thought experiment [91, Page 1068, Chapter 38].

Analysis reveals that the bright bands occur where the path length difference $d \sin \theta$ between the two possible paths equals an integral number of wavelengths λ. Dark bands occur where the path length difference equals a half-odd number of wavelengths. The wavelength is related to the particle momentum p by the deBroglie relation

$$p = \frac{h}{\lambda}, \tag{3.229}$$

where h is Planck's constant. According to Einstein's hypothesis, light propagates in the form of discrete energy packets called *photons*, where each photon obeys this same relationship between momentum and wavelength. Indeed, the same two-slit interference was observed much earlier for light by Young. This experiment and many related topics are authoritatively described by Born and Wolf [11]. This is one of many illustrations of the close correspondence between light optics and particle optics.

As a related intuitive concept, we next consider the propagation of a wave front through space and time, as described by Huygens' principle. This will prove to be indispensible to formulating a mathematical description of diffraction, which is derived in the following sections. It will not be necessary to specifically invoke the discreteness of particles. Rather we will take a more traditional approach, regarding the wave function as continuous in space and time. We will develop a scalar theory, where the optical disturbance is adequately described by the scalar wave function. This is permissible, because we consider only particle motion in a vacuum, which is inherently isotropic. We will ignore the intrinsic spin, since it is not needed for this discussion. The reader is referred in the following to two definitive texts by Born and Wolf [11], and by Goodman [36] for detailed and comprehensive discussion.

3.3.1 The Fresnel–Kirchhoff relation

In mathematical terms, the central problem in diffraction theory is to calculate the amplitude $u(\mathbf{x})$, given specified, known boundary conditions. For a free particle this is a solution to the Helmholtz equation, given by

$$(\nabla^2 + k^2)\, u(\mathbf{x}) = 0, \qquad (3.230)$$

where k is a constant, and $u(\mathbf{x})$ is the spatial part of the wave function. This is precisely the scalar wave equation applicable to light, in which case $k = \omega/c$, and c is the speed of light. Allowing for this, we therefore anticipate that the results to follow are otherwise equally valid for a photon and a charged particle. In this section, we describe a Green's function approach originally derived by Sommerfeld [85] for light optics to achieve this. This methodology is known as the Rayleigh–Sommerfeld solution. The reader is referred to the text by Goodman [36] for a comprehensive discussion, including the interesting historical attempts to correctly understand this problem.

For the present purpose, we assume the particle propagates freely, in the absence of electric and magnetic fields. We begin by stating a very general result, which will prove to be useful. We assume two arbitrary, complex functions $U(\mathbf{x})$ and $V(\mathbf{x})$, where these functions are finite and differentiable over an arbitrary, closed volume τ. We form the quantity $U\,\nabla^2 V - V\,\nabla^2 U$, and integrate this over the volume τ. It follows that

$$\int_\tau \left[U\,\nabla^2 V - V\,\nabla^2 U \right] d\tau = \int_\tau \nabla \cdot \left[U\,\nabla V - V\,\nabla U \right] d\tau. \quad (3.231)$$

We now make use of the fact that, for any vector field $\mathbf{C}(\mathbf{x})$

$$\int_\tau \nabla \cdot \mathbf{C}\, d\tau = \int_S \mathbf{C} \cdot d\mathbf{S}. \tag{3.232}$$

This expresses the fact that the volume integral of the divergence of $\mathbf{C}$ is equivalent to the integral of the outward normal component of $\mathbf{C}$ over the surface $\mathbf{S}$ enclosing the volume τ. This general result is called the divergence theorem. Applying this to the present problem, we find

$$\int_\tau \left[U(\mathbf{x})\, \nabla^2 V(\mathbf{x}) - V(\mathbf{x})\, \nabla^2 U(\mathbf{x}) \right]\, d\tau$$

$$= \int_S \left[U(\mathbf{x})\, \frac{\partial}{\partial n} V(\mathbf{x}) - V(\mathbf{x})\, \frac{\partial}{\partial n} U(\mathbf{x}) \right]\, dS, \tag{3.233}$$

where the right side is the surface integral over the surface S enclosing the volume τ. The quantity n represents the coordinate along a direction locally perpendicular to the surface S, oriented outward from the volume τ. The partial derivative with respect to n is thus the normal gradient of the function.

The relationship (3.233) between the volume and surface integrals is called *Green's theorem*. As the functions U and V are arbitrary, this result is quite general. We will now proceed to apply it to the present problem.

First we consider the special case where $u(\mathbf{x})$ depends only on the magnitude $r = |\mathbf{x}|$. In spherical coordinates, the Helmholtz equation is

$$\frac{1}{r} \frac{d^2}{dr^2} [r\, u(r)] + k^2 u(r) = 0. \tag{3.234}$$

This is integrated immediately to give

$$r\, u(r) = \exp\left(\pm ikr \right), \qquad u(r) = \frac{1}{r} \exp\left(\pm ikr \right), \tag{3.235}$$

which represents a spherical wave about the origin $r = 0$. Multiplying this by $\exp(-i\omega t)$, it is evident that the positive exponent represents an outgoing spherical wave, while the negative exponent represents an incoming spherical wave. The positive and negative exponentials represent two linearly independent solutions to the Helmholtz equation, as required for any second order, ordinary differential equation.

We assume an opaque, planar screen of infinite extent, with one or more apertures or openings of arbitrary shape and position in the screen. This is shown schematically in Figure 3.9. The screen

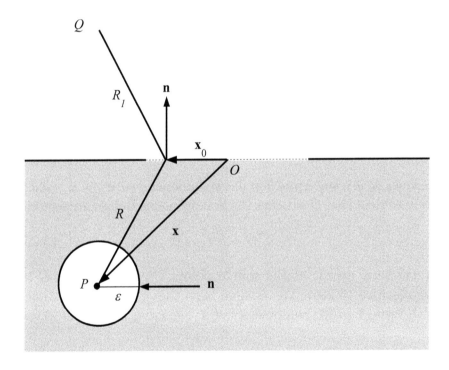

Figure 3.9: Geometry for Sommerfeld's solution by Green's function.

is assumed to be illuminated by an arbitrary collection of sources (not shown), such that the amplitude at position $\mathbf{x}_0$ in the plane of the screen is $u(\mathbf{x}_0)$. This amplitude is assumed to be known.

Given this information, we wish to evaluate the amplitude $u(\mathbf{x})$ at a remote observation point at position $\mathbf{x}$. This point is designated by the point P in the figure. This represents the statement of our problem in mathematical terms.

Following Sommerfeld [85], we postulate two point sources at P and Q, on opposite sides of the screen, and equidistant from it. We imagine a spherical wave emanating separately from each of the two points P and Q, where the two waves are *assumed* to radiate exactly 180 degrees out of phase relative to one another. The resultant amplitude G is found by algebraically adding the complex amplitudes corresponding to the two spherical waves (3.235). This yields

$$G = \frac{1}{R}\exp(ikR) - \frac{1}{R_1}\exp(ikR_1), \qquad (3.236)$$

where the radii R and R_1 are shown in the figure. In the plane of the screen, $R = R_1$, and consequently, $G = 0$. This will be crucially important in the following.

Because G is a superposition of two spherical waves, it is immediately evident that G satisfies the homogeneous Helmholtz equation

$$\nabla^2 G + k^2 G = 0 \qquad (3.237)$$

everywhere, except at the source points P and Q, where G has singularities (3.236). We assume here that the differentiation is with respect to the components of $\mathbf{x}$, shown in Figure 3.9.

We can now write

$$\int_\tau \left[u(\mathbf{x})\,\nabla^2 G(\mathbf{x},\mathbf{x}_0) - G(\mathbf{x},\mathbf{x}_0)\,\nabla^2 u(\mathbf{x}) \right]\,d\tau$$

$$= \int_S \left[u(\mathbf{x})\,\frac{\partial}{\partial n}G(\mathbf{x},\mathbf{x}_0) - G(\mathbf{x},\mathbf{x}_0)\,\frac{\partial}{\partial n}u(\mathbf{x}) \right]\,dS, \quad (3.238)$$

where the left side is an integral over an enclosed volume τ, and the right side is an integral over the surface S enclosing the volume τ. We have made direct use of Green's theorem (3.233), where we have substituted u for U, and G for V. The volume τ is depicted by the shaded area in the figure. A small sphere about the point P is specifically *excluded* from τ, as G has a singularity at P. The closed surface S includes the infinite planar screen, the small sphere about P, and is closed by a hemispherical surface at infinity in the lower half-space of the figure.

The function G defined in (3.236) is called the *Green's function* for this problem. The actual specification of G in (3.236) is not unique, as any well-behaved function G would satisfy Green's theorem (3.233). In practice, the choice of G, together with its boundary conditions, is intentionally made in a way which leads to a simplification of the problem at hand, as the following will show.

Because both u and G satisfy the Helmholtz equation, it follows immediately that the integrand on the left side of (3.238), and hence the left side itself, is identically zero. It should also be added that the sources at P and Q are not physical sources. Rather, they are merely a mathematical construct to aid in solving for u.

The task remains to evaluate the surface integral over S on the right side of (3.238). This is equal to the sum of three individual surface integrals over the hemispherical surface at infinity, the small spherical surface of radius ϵ, and the planar screen, respectively. Considering first the hemispherical surface S_1 at infinity, it is straightforward to show that

$$\int_{S_1} \left[u(\mathbf{x}) \frac{\partial}{\partial n} G(\mathbf{x}, \mathbf{x}_0) - G(\mathbf{x}, \mathbf{x}_0) \frac{\partial}{\partial n} u(\mathbf{x}) \right] dS$$

$$\rightarrow \int_{S_1} \left(iku - \frac{\partial u}{\partial n} \right) GR^2 \, d\Omega, \qquad (3.239)$$

where $d\Omega$ is the solid angle element. As GR is bounded as $R \rightarrow \infty$, it follows that the right side vanishes, as long as

$$R \left(iku - \frac{\partial u}{\partial n} \right) \rightarrow 0 \qquad (3.240)$$

as $R \to \infty$. This is, in fact, the case for a purely outgoing spherical wave. This is known as the Sommerfeld radiation condition. The surface integral over the hemispherical surface S_1 at infinity is zero. Thus, it makes no contribution to the overall surface integral over S, which is the right side of (3.238).

Next, we consider the small spherical surface S_2 of radius ϵ about P. As $\epsilon \to 0$, the integrand on the right side of (3.238) becomes dominated by the first term in the expression (3.236) for the Green's function G. It is straightforward to show that

$$\int_{S_2} \left[u(\mathbf{x}) \frac{\partial}{\partial n} G(\mathbf{x}, \mathbf{x}_0) - G(\mathbf{x}, \mathbf{x}_0) \frac{\partial}{\partial n} u(\mathbf{x}) \right] dS \to -4\pi u(\mathbf{x})$$

$$(3.241)$$

in the limit $\epsilon \to 0$.

Finally, we consider the surface S_0 of the planar screen. We assume $u = 0$ on the interior of the opaque portion of the screen, i.e., the screen is perfectly opaque. We further notice by symmetry that $R = R_1$ everywhere in the plane of the screen. It follows (3.236) that $G = 0$ over the entire plane of the screen. In fact, this is the reason for Sommerfeld's choice of the two equidistant point sources at P and Q, radiating directly out of phase. The two spherical waves from the point sources at P and Q thus interfere destructively at the plane of the screen. This leads to a considerable simplification in the evaluation of the right side of (3.238), by eliminating the second term in the integrand. Considering the surfaces S_0 (the screen) and S_2 (the small sphere) together, it follows (3.241) that

$$4\pi u(\mathbf{x}) = -\int_{S_0} dS_0 \, u(\mathbf{x}_0) \frac{\partial}{\partial n} G(\mathbf{x}, \mathbf{x}_0),$$

$$(3.242)$$

where

$$\frac{\partial G}{\partial n} = \frac{\partial G}{\partial R} \frac{\partial R}{\partial n} + \frac{\partial G}{\partial R_1} \frac{\partial R_1}{\partial n}$$

$$= 2 \cos(n, R) \frac{ikR - 1}{R^2} \exp(ikR) \qquad (3.243)$$

remembering that $R = R_1$. In the limit of short wavelength we have $kR \gg 1$, in which case we can approximate

$$u(\mathbf{r}, z) = \frac{1}{i\lambda} \int d^2\mathbf{r}_0 \, u_0(\mathbf{r}_0, z_0) \, \frac{\exp\left(ik\,|\mathbf{x} - \mathbf{x}_0|\right)}{|\mathbf{x} - \mathbf{x}_0|} \cos\left(\hat{\mathbf{n}}, \mathbf{x} - \mathbf{x}_0\right),$$

$$(3.244)$$

where we have substituted $R = |\mathbf{x} - \mathbf{x}_0|$. Also, $k = 2\pi/\lambda$, where λ is the particle wavelength, given by $\lambda = h/p$. The integral (3.244) need only be calculated over the open areas in the screen, where $u_0(\mathbf{r}_0, z_0)$ is nonzero. Here we have expressed the three-vector position $\mathbf{x}$ as a two-vector position $\mathbf{r}$ in the transverse plane, and an axial position z; i.e., $\mathbf{x} = (\mathbf{r}, z)$. We will continue to use this notation throughout.

The relation (3.244) is known as the Fresnel–Kirchhoff relation for historical reasons. It is a general solution of the Helmholtz equation, expressed in integral form. It represents an approximation, which is only valid in the limit where the wavelength $\lambda \ll R$, that is, the wavelength is small compared with the viewing distance. Within this approximation, the specification of $u(\mathbf{r}_0, z_0)$ is quite general. In practice, it depends on the distribution of physical sources behind the screen. In the special case where the screen is uniformly illuminated at normal incidence from behind by a monochromatic plane wave, $u(\mathbf{r}_0, z_0)$ is independent of $\mathbf{r}_0$, and comes outside the integral as a leading factor.

The integrand in (3.244) includes an outgoing spherical wave emanating from the point $\mathbf{x}_0$. The integral represents a coherent summation of all spherical waves emanating from within the aperture. Physically, this is an expression of Huygens' principle. This in turn determines the downstream amplitude $u(\mathbf{r}, z)$, given a known amplitude $u_0(\mathbf{r}_0, z_0)$ in the plane of the screen z_0. The intensity in the plane z is then given by $|u(\mathbf{r}, z)|^2$. We will see in the following sections that the Fresnel–Kirchhoff equation (3.244) can be used in a very practical way to understand the intensity distribution for a rich variety of configurations.

Problem

Show by direct substitution that the solution (3.244) satisfies the
time-independent wave equation (3.230).

3.3.2   The Fresnel and Fraunhofer approximations

The Fresnel–Kirchhoff relation (3.244), is amenable to numerical integration to obtain an exact expression for the amplitude
$u(\mathbf{r}, z)$. In this section we make several approximations which will
permit straightforward analytical evaluation of the integral. This
approach allows a more direct physical insight for a large variety
of interesting cases. Assuming small angles, the ray slope is much
less than unity, in which case

$$\cos\left(\hat{\mathbf{n}}, \mathbf{x} - \mathbf{x}_0\right) \approx 1. \tag{3.245}$$

We further adopt the simplifying approximation

$$|\mathbf{x} - \mathbf{x}_0| = \sqrt{(\mathbf{r} - \mathbf{r}_0)^2 + Z^2} \approx Z + \frac{1}{2Z}\left(\mathbf{r}^2 + \mathbf{r}_0^2 - 2\mathbf{r}\cdot\mathbf{r}_0\right), \tag{3.246}$$

where $Z = z - z_0$ is the drift length. This is often referred to
as the *parabolic* approximation, as the spherical wavefront is approximated by a parabolic surface for small angles. With these
approximations, (3.244) reduces to

$$u(\mathbf{r}, z) = \frac{1}{i\lambda Z} \exp\left[ik\left(Z + \frac{\mathbf{r}^2}{2Z}\right)\right]$$
$$\cdot \int d^2\mathbf{r}_0\, u_0(\mathbf{r}_0, z_0) \exp\left[\frac{ik}{Z}\left(\frac{\mathbf{r}_0^2}{2} - \mathbf{r}\cdot\mathbf{r}_0\right)\right]. \tag{3.247}$$

This is known as the *Fresnel* approximation.

Next we investigate the special case where

$$\frac{kr_0^2}{2Z} \ll 2\pi \qquad (3.248)$$

at *all* positions $\mathbf{r}_0$. Mathematically, the phase shift due to the first term in the exponent is negligible.

Recalling that $k = 2\pi/\lambda$, this is equivalent to

$$\frac{r_0^2}{2Z} \ll \lambda, \qquad (3.249)$$

where r_0/Z is the tangent of the angle subtended on the central axis at the end plane. It follows that the first term in the exponent can be ignored. In this case (3.247) reduces to

$$
\begin{aligned}
u(\mathbf{r}, z) &= \frac{1}{i\lambda Z} \exp\left[ik\left(Z + \frac{\mathbf{r}^2}{2Z}\right)\right] \\
&\quad \cdot \int d^2\mathbf{r}_0 \, u_0(\mathbf{r}_0, z_0) \exp\left(\frac{-ik\,\mathbf{r}\cdot\mathbf{r}_0}{Z}\right).
\end{aligned} \qquad (3.250)
$$

This is referred to as the *Fraunhofer* approximation. This approximation is valid for Z sufficiently large, that is, the observation plane is sufficiently far removed from the plane of the screen. We see from (3.250) that the amplitude $u(\mathbf{r}, z)$ is proportional to the Fourier transform of $u_0(\mathbf{r}_0, z_0)$ with the transform variable $k\mathbf{r}/Z$, where r/Z is the tangent of the viewing angle in the observation plane.

The intensity is given by the absolute square of $u(\mathbf{r}, z)$. The leading phase factor in (3.247, 3.250) drops out in the expression for the intensity, and can therefore be ignored. The intensity is directly measurable, whereas the amplitude $u(\mathbf{r}, z)$ is not. The amplitude can only be deduced by measuring the intensity in an interference experiment, where the relative phase of the interfering waves is precisely known. We therefore ascribe direct physical significance to the intensity, but not the amplitude.

We have derived a transformation of the wave function $u(\mathbf{r}, z)$ between successive planes in the drift length of an optical system. To this point we have not assumed any particular symmetry, Cartesian, axial, or otherwise. In the following, we will assume axial symmetry. This simplifying assumption is applicable to many practical systems.

Next, we wish to incorporate the focusing effects of a lens. This is depicted in Figure 3.10, where a thin lens is located at the plane

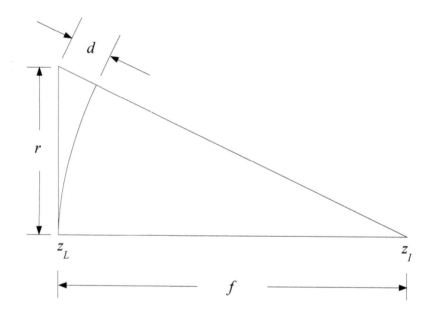

Figure 3.10: Path length shift for a thin lens.

z_L. Ideally, rays at all radii r focus to a common point in the plane z_I. This ideal focusing only occurs for rays close to the optic axis. We therefore refer to this ideal focusing as the paraxial approximation. Considering the extreme ray, we see by striking a circular arc that its path length is longer than the axial ray by a distance d. The circular arc coincides with a surface of constant phase for a wave converging to the image point. From the Pythagorean theorem,

$$r^2 + f^2 = (f + d)^2 \approx f^2 + 2fd, \tag{3.251}$$

where we assume $d \ll 2f$. In this approximation we have

$$d = \frac{r^2}{2f}. \tag{3.252}$$

This gives rise to a phase shift $-kd$ at the plane of the thin lens z_L. Equivalently, the wave function is multiplied by a phase factor given by

$$L_f(r) = \exp\left(\frac{-ikr^2}{2f}\right) \tag{3.253}$$

for the paraxial approximation (no aberration).

Using these transformations, we can build up a simple optical system. We apply successive transformations, first for the object space, followed by the lens, and finally followed by the image space. We define

$$
\begin{aligned}
z_0 &= \text{object plane} \\
z_1 &= \text{lens plane} \\
z &= \text{recording plane} \\
Z_1 &= z_1 - z_0 = \text{object distance} \\
Z_2 &= z - z_1 = \text{image distance.} \tag{3.254}
\end{aligned}
$$

We further denote $\mathbf{r}_0, \mathbf{r}_1$, and $\mathbf{r}$ as the two-dimensional position vectors in the object, lens, and recording planes, respectively.

We assume a pupil located at the lens plane z_1. By successive transformations, interchanging the order of integrations, it is straightforward to show that

$$
\begin{aligned}
u(\mathbf{r}, z) &= \frac{-1}{\lambda^2 Z_1 Z_2} \exp\left[ik\left(Z_1 + Z_2 + \frac{\mathbf{r}^2}{2Z_2}\right)\right] \\
&\quad \cdot \int d^2\mathbf{r}_0 \, u_0(\mathbf{r}_0, z_0) \exp\left(\frac{ik\mathbf{r}_0^2}{2Z_1}\right) h(\mathbf{r}_0, \mathbf{r}), \tag{3.255}
\end{aligned}
$$

where we have defined a kernel h given by

$$
\begin{aligned}
h(\mathbf{r}_0, \mathbf{r}) &= \int d^2\mathbf{r}_1\, P(\mathbf{r}_1)\, \exp\left[ik\,\frac{\mathbf{r}_1^2}{2}\left(\frac{1}{Z_1}+\frac{1}{Z_2}-\frac{1}{f}\right)\right] \\
&\quad \cdot\; \exp\left[-ik\mathbf{r}_1\cdot\left(\frac{\mathbf{r}_0}{Z_1}+\frac{\mathbf{r}}{Z_2}\right)\right],
\end{aligned}
$$

$$(3.256)$$

where $P(\mathbf{r}_1)$ is the pupil transmission function, equal to unity in the transmitting area, and zero otherwise.

In the special case where P represents a round aperture centered on the optic axis, it is advantageous to use polar coordinates $\mathbf{r} = (\rho, \phi)$. We perform the azimuthal integral first, where J_0 is the zero-order Bessel function with integral representation given by

$$
J_0(x) = \frac{1}{2\pi}\int_0^{2\pi} e^{-ix\cos\phi}\, d\phi.
$$

$$(3.257)$$

From this it follows immediately that

$$
\begin{aligned}
h(\mathbf{r}_0, \mathbf{r}) &= 2\pi \int_0^{\infty} d\rho_1\, \rho_1\, P(\rho_1)\, \exp\left[ik\,\frac{\rho_1^2}{2}\left(\frac{1}{Z_1}+\frac{1}{Z_2}-\frac{1}{f}\right)\right] \\
&\quad \cdot\; J_0\left(k\rho_1\left|\frac{\mathbf{r}_0}{Z_1}+\frac{\mathbf{r}}{Z_2}\right|\right).
\end{aligned}
$$

$$(3.258)$$

This gives a general expression for the optical transformation, independent of the location of the start and end planes relative to the focal plane of the lens. In the following sections, we apply this to several important special cases.

Problems

1. Show that for an ideal point object, a transformation (3.250) from z_0 to z followed by a second transformation from z to z_2 is equivalent to a single transformation from z_0 to z_2.

2. An electron with kinetic energy 100 keV is normally incident on a circular object of radius 100 nm. Estimate the shortest distance Z for which the Fraunhofer approximation is a valid estimate of the downstream amplitude.

3.3.3 Amplitude in the Gaussian image plane

We assume an object plane at axial coordinate z_O, a thin lens of focal length f at z_L, and a Gaussian image plane at z_I. This geometry is shown in Figure 3.11. The relationship between the

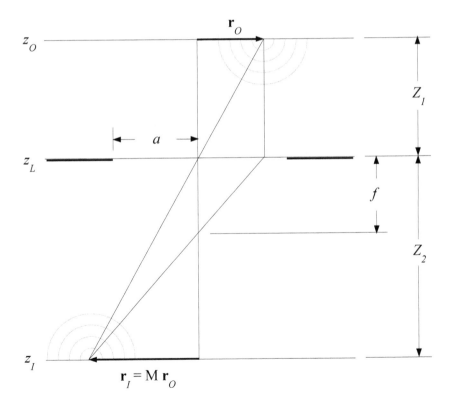

Figure 3.11: Image formation for a point object at $\mathbf{r}_O$.

object distance Z_1, the image distance Z_2, and the lens focal length f is given for ideal imaging as

$$\frac{1}{Z_1} + \frac{1}{Z_2} - \frac{1}{f} = 0, \qquad \mathrm{M} = -\frac{Z_2}{Z_1}, \qquad (3.259)$$

where M is the lateral magnification. An outgoing spherical wave emanates from the object point at lateral position $\mathbf{r}_O$. In the limit of perfect imaging (paraxial approximation), an incoming spherical wave converges on the conjugate image point $\mathbf{r}_I = \mathrm{M}\mathbf{r}_O$. We assume a round aperture of radius a coplanar with the lens at z_L. The pupil function $P(\mathbf{r}_1)$ is unity for $0 \leq r_1 \leq a$, and zero for $r_1 > a$. Inserting 3.259 directly into the kernel h in 3.258, we obtain

$$h(\mathbf{r}_O, \mathbf{r}_I) = 2\pi \int_0^a dr_1 \, r_1 \, J_0 \left(\frac{kr_1 \, |\mathbf{r}_I - \mathrm{M}\mathbf{r}_O|}{Z_2} \right). \qquad (3.260)$$

This is recognizable as the Bessel transform of the pupil function. From this it follows immediately that

$$h(\mathbf{r}_O, \mathbf{r}_I) = 2\pi a^2 \left(\frac{ka|\mathbf{r}_I - \mathrm{M}\mathbf{r}_O|}{Z_2} \right)^{-1} J_1 \left(\frac{ka|\mathbf{r}_I - \mathrm{M}\mathbf{r}_O|}{Z_2} \right), \qquad (3.261)$$

where we have made use of the integral

$$\int J_0(x) \, x \, dx = x \, J_1(x). \qquad (3.262)$$

Substituting in 3.255, we obtain the amplitude in the Gaussian image plane $z = z_I$ as

$$
\begin{aligned}
u_I(\mathbf{r}_I) \;=\; & \frac{k^2 a^2}{2\pi Z_1 Z_2} \int d^2 \mathbf{r}_O \, u_O(\mathbf{r}_O, z_O) \exp \left(\frac{ik\mathbf{r}_O^2}{2Z_2} \right) \\
\cdot \; & \left(\frac{ka|\mathbf{r}_I - \mathrm{M}\mathbf{r}_O|}{Z_2} \right)^{-1} J_1 \left(\frac{ka|\mathbf{r}_I - \mathrm{M}\mathbf{r}_O|}{Z_2} \right), \quad (3.263)
\end{aligned}
$$

where we have ignored leading phase factors outside the integral, as such factors do not affect the intensity $|u_I(\mathbf{r}_I)|^2$. The complex amplitude u_O represents an extended object. Every object point

$\mathbf{r}_O$ can be considered to be the source of a spherical outgoing wave. The waves emanating from neighboring object points are assumed to radiate coherently with respect to one another. This can only happen if all object points radiate monochromatically with a constant phase relationship. It therefore represents an approximation.

The waves from all object points propagate coherently through the optical system. The integral over $\mathbf{r}_O$ represents a superposition of amplitudes over the entire object plane. The complex amplitude u_O in the object plane is convolved with the function h to form the amplitude u_I in the Gaussian image plane. This physical significance of this can be appreciated by considering the important special case of a point object on axis. In this case the amplitude in the object plane is given by

$$u_O(\mathbf{r}_O) = \delta(\mathbf{r}_O), \tag{3.264}$$

where the right-hand side is the Dirac delta function. From the property of the delta function, it follows immediately that

$$u_I(\mathbf{r}_I) \sim \left(\frac{kar_I}{Z_2}\right)^{-1} J_1\left(\frac{kar_I}{Z_2}\right), \tag{3.265}$$

where the ratio a/Z_2 is the tangent of the semiangle of the cone of rays at the image plane z_I. The square of this functional form, which represents the intensity, is known as an Airy disk. Physically, this is precisely the diffraction pattern of the aperture. The kernel h is called the *point spread function*, since it represents the blurring of every image point relative to an ideal image.

To this point we have assumed imaging without aberrations. We now inquire into the effect of spherical aberration. This is depicted in Figure 3.12, where the spherical aberration gives rise to an additional path length increment d_S. The spherical aberration in the Gaussian image plane was found earlier to be

$$\delta r_S = C_S \alpha^3, \tag{3.266}$$

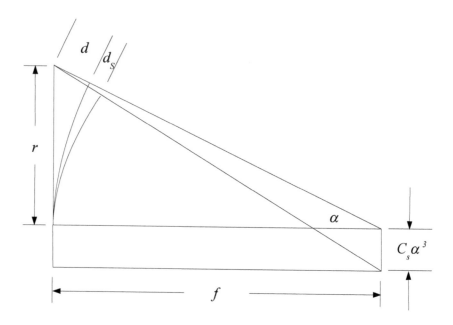

Figure 3.12: Path length shift for a thin lens with spherical aberration.

where α is the semiangle made by the extreme ray with the optic axis. Applying the Pythagorean theorem,

$$(r + \delta r_S)^2 + f^2 = (f + d + d_S)^2 , \qquad (3.267)$$

where we wish to solve for the path length increment d_S. For small angles, this is approximated by

$$d_S \approx C_S \left(\frac{r}{f}\right)^4 . \qquad (3.268)$$

In this approximation, the resulting phase shift is $-kd_S$.

We next investigate the behavior of the wave function in a plane which is slightly displaced from the Gaussian image plane by a defocus distance δf. Recalling the earlier expression for the change of path length d for a thin lens of focal length f, we replace f by

$f + \delta f$. Retaining only terms through first order in δf, this leads to a path length increment due to defocus given by

$$d_f = \frac{r^2 (\delta f)}{2 f^2}, \qquad (3.269)$$

where this in turn leads to a phase shift $-k d_f$. Taking spherical aberration and defocus into account, the expression 3.253 for a thin lens is modified as a multiplicative phase factor given by

$$L_f(\mathbf{r}) = \exp\left[\frac{-ikr^2}{2f} \left(1 + \frac{\delta f}{f} + \frac{2C_S r^2}{f^3} \right) \right]. \qquad (3.270)$$

Substituting this phase factor into the kernel $h(\mathbf{r}_O, \mathbf{r}_I)$, we obtain the modified expression for the case with spherical aberration and defocus present,

$$\begin{aligned} h(\mathbf{r}_O, \mathbf{r}_I) \;=\; & 2\pi \int_0^a dr_1\, r_1 \exp\left[\frac{-ikr_1^2}{2f} \left(\frac{\delta f}{f} + \frac{2C_S r_1^2}{f^3} \right) \right] \\ & \cdot\; J_0\left(\frac{kr_1 |\mathbf{r}_I - M\mathbf{r}_O|}{Z_2} \right). \end{aligned}$$

$$(3.271)$$

The resulting complex wave function in the Gaussian image plane is

$$\begin{aligned} u_I(\mathbf{r}_I) \;=\; & \frac{-1}{\lambda^2 Z_1 Z_2} \exp\left[ik \left(Z_1 + Z_2 \right) \right] \\ & \cdot\; \int d^2 \mathbf{r}_O\, u_O(\mathbf{r}_O) \exp\left(\frac{ikr_O^2}{2Z_1} \right) h(\mathbf{r}_O, \mathbf{r}_I), \quad (3.272) \end{aligned}$$

recalling that $\lambda = 2\pi/k$. The leading phase factor can be ignored, since it does not appear in the intensity $|u_I|^2$. The phase factor under the integral approaches unity for $kr_O^2/(2Z_1) \ll 2\pi$. However, one must exercise caution before making this approximation for an energetic charged particle, since the wave number k is often quite large.

3.3.4   Amplitude in the diffraction plane

We now consider the special case where

$$Z_1 = Z_2 = f. \tag{3.273}$$

This is shown schematically in Figure 3.13. We see that a given ray slope in the object plane z_O is mapped to a specific, single transverse position in the diffraction plane z_D, regardless of transverse position in the object plane. The diffraction plane z_D is the plane where a diffraction pattern of a periodic object comes into sharp focus.

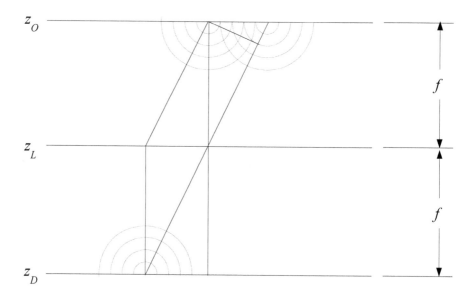

Figure 3.13: Formation of a diffraction pattern.

Evaluating the kernel h in (3.258) between the object plane $z = z_O$ and the diffraction plane $z = z_D$, we find

$$h(\mathbf{r}_O, \mathbf{r}_D) = 2\pi \int_0^\infty dr_1 \, r_1 \, \exp\left(\frac{ikr_1^2}{2f}\right) J_0\left(\frac{kr_1}{f}|\mathbf{r}_O + \mathbf{r}_D|\right). \tag{3.274}$$

We have assumed no aperture, in which case $P(\mathbf{r}_1) = 1$ for all $\mathbf{r}_1$. It follows that

$$h(\mathbf{r}_O, \mathbf{r}_D) = i\lambda f \exp\left[-\frac{ik}{2f}(\mathbf{r}_O + \mathbf{r}_D)^2\right], \tag{3.275}$$

where we have made use of the integral

$$\int_0^\infty \exp\left(i\alpha x^2\right) J_0(\beta x)\, x\, dx = \frac{i}{2\alpha} \exp\left(-\frac{i\beta^2}{4\alpha}\right), \qquad (\alpha \neq 0). \tag{3.276}$$

The amplitude $u_D(z_D)$ in the diffraction plane is given by

$$\begin{aligned}
u_D(\mathbf{r}_D) &= \frac{1}{i\lambda f} \exp\left[ik\left(2f + \frac{\mathbf{r}_D^2}{2f}\right)\right] \cdot \int d^2\mathbf{r}_O\, u_O(\mathbf{r}_O) \\
&\quad \cdot \exp\left(\frac{ik\mathbf{r}_O^2}{2f}\right) \exp\left[-\frac{ik}{2f}(\mathbf{r}_O + \mathbf{r}_D)^2\right]. \tag{3.277}
\end{aligned}$$

Expanding,

$$(\mathbf{r}_O + \mathbf{r}_D)^2 = \mathbf{r}_O^2 + \mathbf{r}_D^2 + 2\,\mathbf{r}_O \cdot \mathbf{r}_D. \tag{3.278}$$

Substituting into 3.277, it follows that the amplitude in the diffraction plane is given by

$$u_D(\mathbf{r}_D) = \frac{1}{i\lambda f} \int d^2\mathbf{r}_O\, u_O(\mathbf{r}_O) \exp\left(-\frac{ik\mathbf{r}_O \cdot \mathbf{r}_D}{f}\right), \tag{3.279}$$

ignoring the leading exponential phase factor, as this does not influence the intensity $|u_D|^2$. This is recognizable as a Fourier transform of the object, with the transform variable $k\mathbf{r}_D/f$. For this reason, the diffraction plane z_D is often referred to as the *Fourier plane*.

Geometrically, each specific value of the ray slope in the object plane is mapped into a unique position in the Fourier plane. This enables one to directly obtain an intensity map of a diffraction pattern. We notice that $\mathbf{r}_D/f$ is the ray slope at the object. Equivalently, this is the tangent of the diffraction angle.

Constructive interference occurs when the path difference between neighboring rays is an integral number of wavelengths. This can be seen in the figure, where the surfaces of constant phase form concentric spheres centered on each object point. All object points are assumed to radiate coherently with respect to each other.

3.3.5 Optical transformation for a general imaging system with coherent illumination

In the preceding analysis, we obtained the optical transformation for a simple system in two specific configurations, each employing of a single lens with focal length f. This was done for the formation of an image, and separately, formation of a diffraction pattern. We assumed perfectly coherent illumination, where a constant phase relationship exists between all points of the object, and the illumination is monochromatic. The basic optical elements are a drift length, a thin lens, and a pupil. In principle, these can be applied in any order, and to any degree of complexity, to build up an arbitrary optical system. Given these mathematical tools, we are now in a position to consider a general imaging system, consisting of an arbitrary configuration of optical elements. In this section and the next, we closely follow the analysis of Goodman [36].

We require only that an image be formed. Here the object is represented by a complex amplitude $u_O(\mathbf{r}_O)$, and the image is represented by a complex amplitude $u_I(\mathbf{r}_I)$. The relevant qantity of physical interest is $|u_I(\mathbf{r}_I)|^2$, which is the intensity measured in the image plane z_I. We seek an optical transformation which expresses u_I in terms of u_O. Such a transformation would contain all of the physically relevant information about the quality of the image; i.e., how closely the image approximates an ideal replica of the object. In order to be useful, this procedure must account for aberrations, defocus, and diffraction with a finite pupil, all of which tend to degrade the image.

We begin our analysis by considering a specific hypothetical optical configuration, consisting of two ideal lenses of focal lengths f_1 and f_2, respectively. This is shown schematically in Figure 3.17. By assumption, a physical aperture is located in the Fourier plane z_A of the first lens. Furthermore, the image plane is assumed to lie in the Fourier plane of the second lens. The back focal plane of the first lens thus coincides with the front focal plane of the second lens. By inspection, it is easy to see that a real image of the original object in the plane z_O is formed in the plane z_I. The magnification is given by

$$M = -\frac{f_2}{f_1},\qquad(3.280)$$

where the minus sign indicates that the image is inverted with respect to the object. From (3.250) the amplitude $u_A(\mathbf{r}_A)$ is expressed in terms of the object $u_O(\mathbf{r}_O)$ in the Fraunhofer approximation by

$$u_A(\mathbf{r}_A) = \frac{1}{i\,\lambda\,f_1}\int d^2\mathbf{r}_O\,u_O(\mathbf{r}_O)\,\exp\left(-\frac{ik\,\mathbf{r}_O\cdot\mathbf{r}_A}{f_1}\right).\qquad(3.281)$$

Similarly, the amplitude $u_I(\mathbf{r}_I)$ is expressed in terms of $u_A(\mathbf{r}_A)$ by

$$u_I(\mathbf{r}_I) = \frac{1}{i\,\lambda\,f_2}\int d^2\mathbf{r}_A\,u_A(\mathbf{r}_A)\,P(\mathbf{r}_A)\,\exp\left(-\frac{ik\,\mathbf{r}_A\cdot\mathbf{r}_I}{f_2}\right),\qquad(3.282)$$

where $P(\mathbf{r}_A)$ is the pupil function. Substituting the first of these equations into the second, and interchanging the order of integrations, we obtain

$$u_I(\mathbf{r}_I) = \frac{-1}{\lambda^2\,f_1\,f_2}\int d^2\mathbf{r}_O\,u_O(\mathbf{r}_O)\int d^2\mathbf{r}_A\,P(\mathbf{r}_A)$$
$$\cdot\;\exp\left[-\frac{ik}{f_2}\mathbf{r}_A\cdot(\mathbf{r}_I - M\mathbf{r}_O)\right],\qquad(3.283)$$

where we have made use of the expression for the magnification M above.

We now define a two-vector position $\mathbf{r}_G$ in the image plane z_I as

$$\mathbf{r}_G = \mathrm{M}\,\mathbf{r}_O. \tag{3.284}$$

The position $\mathbf{r}_G$ represents the position $\mathbf{r}_O$ in the object plane z_O transferred to the Gaussian image plane z_I by ideal imaging in the limit of geometrical optics. Furthermore, we define an amplitude $u_G(\mathbf{r}_G)$ in the Gaussian image plane as

$$u_G(\mathbf{r}_G) = \frac{1}{\mathrm{M}}\,u_O(\mathbf{r}_O) = \frac{1}{\mathrm{M}}\,u_O\left(\frac{\mathbf{r}_G}{\mathrm{M}}\right), \tag{3.285}$$

where the object function $u_O(\mathbf{r}_O)$ is assumed to be known. Thus $u_G(\mathbf{r}_G)$ represents the ideal image. By inspection, this preserves the normalization, namely,

$$\int d^2\mathbf{r}_G\,|\,u_G(\mathbf{r}_G)\,|^2 = \int d^2\mathbf{r}_O\,|\,u_O(\mathbf{r}_O)\,|^2, \tag{3.286}$$

where $d^2\mathbf{r}_G = \mathrm{M}^2\,d^2\mathbf{r}_O$. This allows us to write

$$u_I(\mathbf{r}_I) = \int d^2\mathbf{r}_G\,u_G(\mathbf{r}_G)\,H(\mathbf{r}_I - \mathbf{r}_G), \tag{3.287}$$

where we have defined a new kernel H from (3.283, 3.284, 3.285) by

$$H(\mathbf{r}_I - \mathbf{r}_G) = \frac{1}{\lambda^2\,f_2^2}\int d^2\mathbf{r}_A\,P(\mathbf{r}_A)\,\exp\left[-\frac{ik}{f_2}\,\mathbf{r}_A\cdot(\mathbf{r}_I - \mathbf{r}_G)\right]. \tag{3.288}$$

We see from the form of (3.287) that H is a point spread function, and from (3.288) that H is the Fourier transform of the pupil function P.

It is informative to study this in the Fourier space of spatial frequencies. We define the two-dimensional Fourier transforms

$$\tilde{u}_I(\mathbf{K}) = \int d^2\xi \, u_I(\xi) \, e^{-i\mathbf{K}\cdot\xi}$$

$$\tilde{u}_G(\mathbf{K}) = \int d^2\xi \, u_G(\xi) \, e^{-i\mathbf{K}\cdot\xi}$$

$$\tilde{H}(\mathbf{K}) = \int d^2\xi \, H(\xi) \, e^{-i\mathbf{K}\cdot\xi}, \tag{3.289}$$

where $\mathbf{K}$ is the two-vector transform variable, and the integration variable ξ is a two-vector position having mathematical significance, but no particular physical significance. The physical significance of $\mathbf{K}$ can be understood by considering a sinusoidal object, with spatial period Λ_I in the image plane. In this case,

$$K = \frac{2\pi}{\Lambda_I} \tag{3.290}$$

in one Cartesian axis. Thus K is 2π times the spatial frequency $1/\Lambda_I$. Applying the convolution theorem (see Appendix A) to (3.287) it follows immediately that

$$\tilde{u}_I(\mathbf{K}) = \tilde{u}_G(\mathbf{K}) \, \tilde{H}(\mathbf{K}). \tag{3.291}$$

Thus, the spatial frequency spectrum of the ideal image is modulated by $\tilde{H}$ to yield the spatial frequency spectrum of the actual image. For this reason, $\tilde{H}$ is called the *amplitude transfer function* or ATF.

To understand the physical significance of this, we substitute in the expression for the kernel H. This gives

$$\tilde{H}(\mathbf{K}) = \int d^2\xi \, e^{-i\mathbf{K}\cdot\xi} \left[\frac{1}{\lambda^2 f_2^2} \int d^2\mathbf{r}_A \, P(\mathbf{r}_A) \exp\left(-\frac{ik}{f_2} \mathbf{r}_A \cdot \xi \right) \right]. \tag{3.292}$$

Interchanging the order of integrations, this gives

$$\tilde{H}(\mathbf{K}) = \int d^2\mathbf{r}_A \, P(\mathbf{r}_A) \cdot \left\{ \frac{1}{\lambda^2 f_2^2} \int d^2\xi \, \exp\left[-i\xi \cdot \left(\mathbf{K} + \frac{k\mathbf{r}_A}{f_2} \right) \right] \right\}. \tag{3.293}$$

We define a new integration variable η by the substitution

$$\xi = \frac{f_2}{k} \eta, \qquad d^2\xi = \left(\frac{f_2}{k} \right)^2 d^2\eta. \tag{3.294}$$

Substituting, this yields

$$\tilde{H}(\mathbf{K}) = \int d^2\mathbf{r}_A \, P(\mathbf{r}_A) \left\{ \frac{1}{(2\pi)^2} \int d^2\eta \, \exp\left[-i\eta \cdot \left(\mathbf{r}_A + \frac{f_2}{k}\mathbf{K} \right) \right] \right\},$$
$$(3.295)$$

remembering that $k = 2\pi/\lambda$. We recognize the expression in curly brackets as a Dirac delta function in two dimensions, where

$$\delta\left(\mathbf{r}_A + \frac{f_2}{k}\mathbf{K} \right) = \frac{1}{(2\pi)^2} \int d^2\eta \, \exp\left[-i\eta \cdot \left(\mathbf{r}_A + \frac{f_2}{k}\mathbf{K} \right) \right].$$
$$(3.296)$$

By the property of the delta function, we immediately perform the integration over $\mathbf{r}_A$, yielding

$$\tilde{H}(\mathbf{K}) = P\left(-\frac{f_2}{k}\mathbf{K} \right). \qquad\qquad (3.297)$$

Mathematically, the amplitude transfer function $\tilde{H}$ is the scaled pupil function. This result is quite general, in that it applies to any aperture, which can be represented by a pupil function P. In the special case of a round aperture of radius a, we have $P = 0$ for $K > k\,a/f_2$. This value of K represents 2π times a cutoff spatial frequency, above which no information is transmitted. The amplitude transfer function is plotted in Figure 3.14. Physically, the pupil cuts off all diffracted orders with spatial frequency larger than the cutoff frequency. The aperture thus acts as a low-pass filter for spatial frequencies. The absence of high spatial frequencies in the image translates to blur.

With this preparation, we are now in a position to address the response of an arbitrary optical system. To this end we state a key hypothesis, namely, every optical system, however complicated, can be represented for analytical purposes by an equivalent two-lens confocal system shown schematically in Figure 3.17. The confocal system represents the optical transfer of object to image in the paraxial approximation, since both lenses of the confocal system are assumed to be ideal. This system also properly represents the effect of diffraction at the exit pupil. We assume the beam kinetic energy to be constant in the equivalent confocal system, and

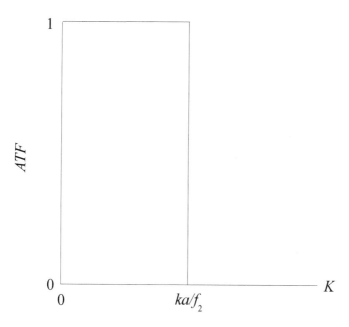

Figure 3.14: Amplitude transfer function, round aperture, paraxial approximation.

equal to the landing energy in the image plane of the real system.

In order for the confocal system to properly represent the real system, we must account for aberrations and defocus. The geometrical aberrations of the real system depend on the coordinates (x_O, y_O) in the object plane, and the coordinates (x_A, y_A) in the aperture plane of the real system. For now, we assume the object coordinates to be fixed, and the aperture coordinates to be variable. It follows that the coordinates (x_G, y_G) of the ideal image are fixed as well. We regard the coordinates (x_I, y_I) in the image plane to be variable. A cone of rays impinges on the image point, where each ray in the cone intersects a unique point (x_A, y_A) in the aperture plane. This is true both in the real system and the equivalent confocal system. Each ray has a unique amount of aberration, which is expressed as an incremental shift δV_{OI} in optical path length of the real system. The primary aberration is represented in the special case of axial symmetry by

$$\delta V_{OI} = \delta \int_{z_O}^{z_I} m\, dz = \int_{z_O}^{z_I} m_4\, dz. \tag{3.298}$$

An explicit expression for this was given earlier in (2.213). At this point we make a key assumption, namely, the shift δV_{OI} is applied discontinuously and entirely in the aperture plane z_A for the equivalent confocal system. This is conceptually equivalent to inserting a phase plate in the aperture plane, where the phase shift varies with coordinates (x_A, y_A) in the equivalent confocal system. Mathematically, we multiply $P(\mathbf{r}_A)$ in (3.282) by a phase factor. This amounts to making the substitution

$$P(\mathbf{r}_A) \rightarrow P(\mathbf{r}_A) \exp\left[\frac{i}{\hbar} \delta V_{OI}(\mathbf{r}_A)\right] \tag{3.299}$$

in the expression (3.288) for $H(\mathbf{r}_I - \mathbf{r}_G)$. The phase shift locally distorts the wave front in the aperture plane of the confocal system. This, in turn deflects the classical ray by a small amount, since the canonical momentum vector is locally normal to the wave front. This results in a lateral displacement of the ray in the image plane, as depicted schematically by the broken lines in Figure 3.17.

Next we inquire into the effect of defocus. We represent this as a small shift of δf in the focal length f_2. We thus make the replacement

$$f_2 \rightarrow f_2 + \delta f = f_2 \left(1 + \frac{\delta f}{f_2}\right). \tag{3.300}$$

Retaining only terms to first order in δf, this leads to the replacement

$$\exp\left[-\frac{ik}{f_2}\mathbf{r}_A \cdot (\mathbf{r}_I - \mathbf{r}_G)\right] \rightarrow$$
$$\exp\left[-\frac{ik}{f_2}\mathbf{r}_A \cdot (\mathbf{r}_I - \mathbf{r}_G)\right] \cdot \exp\left[\frac{ik(\delta f)}{f_2^2}\mathbf{r}_A \cdot (\mathbf{r}_I - \mathbf{r}_G)\right] \tag{3.301}$$

in (3.288). The complete expression for the kernel H is thus given in the presence of aberrations and defocus as

$$
\begin{aligned}
H(\mathbf{r}_I - \mathbf{r}_G) &= \frac{1}{\lambda^2 f_2^2} \int d^2\mathbf{r}_A\, P(\mathbf{r}_A) \exp\left[-\frac{ik}{f_2} \mathbf{r}_A \cdot (\mathbf{r}_I - \mathbf{r}_G) \right] \\
&\cdot \exp\left[\frac{i}{\hbar} \delta V_{OI}(\mathbf{r}_A) + \frac{ik(\delta f)}{f_2^2} \mathbf{r}_A \cdot (\mathbf{r}_I - \mathbf{r}_G) \right],
\end{aligned}
$$
(3.302)

where the integral is performed over the aperture plane of the equivalent confocal system. The aberrations and defocus are contained in the final phase factor on the right side. The amplitude $u_I(\mathbf{r}_I)$ is given by (3.287) with the point spread function H given by (3.302). This provides a quantitative assessment of image fidelity for a general optical system with arbitrary configuration. It thus represents the main result of this section.

3.3.6 Optical transformation for a general imaging system with incoherent illumination

In the preceding section, the illumination was assumed to be coherent. Ideally, this means that the illumination of the object plane is perfectly monochromatic, corresponding to a single eigenstate of definite energy and momentum. It also means that all points in the object plane to radiate with a constant phase relationship to one another. According to the postulates of quantum mechanics, the amplitudes for alternative paths are added in the measurement plane, with the absolute square of the resultant amplitude giving the intensity.

In this section, we consider the case of *incoherent* illumination. By definition, this implies that neighboring object points radiate independently, with relative phase completely uncorrelated. In this case the resulting intensity at the image plane is calculated by adding *intensities* from alternative paths. We define the intensity in the object and image planes respectively as

$$
\begin{aligned}
I_O(\mathbf{r}_O) &= |u_O(\mathbf{r}_O)|^2 \\
I_I(\mathbf{r}_I) &= |u_I(\mathbf{r}_I)|^2.
\end{aligned}
\tag{3.303}
$$

The ideal intensity in the image plane is a perfect magnified replica of the object intensity in the limit of geometrical optics. We define this as

$$
I_G(\mathbf{r}_G) = \frac{1}{M^2} I_O(\mathbf{r}_O) = \frac{1}{M^2} I_O\left(\frac{\mathbf{r}_G}{M}\right).
\tag{3.304}
$$

From (3.287), and the fact that $I_I(\mathbf{r}_I) = |u_I(\mathbf{r}_I)|^2$ we obtain

$$
\begin{aligned}
I_I(\mathbf{r}_I) &= \left[\int d^2 r_G\, u_G(\mathbf{r}_G)\, H(\mathbf{r}_I - \mathbf{r}_G)\right] \\
&\cdot \left[\int d^2 r_G'\, u_G^*(\mathbf{r}_G')\, H^*(\mathbf{r}_I - \mathbf{r}_G')\right],
\end{aligned}
\tag{3.305}
$$

where $\mathbf{r}_G = M\mathbf{r}_O$ is the object point transferred to the Gaussian image plane by ideal imaging. At this point we make a key assumption, namely, that total incoherence implies that only points where $\mathbf{r}_G = \mathbf{r}_G'$ contribute to the result. Mathematically, this is equivalent to inserting a delta function $\delta(\mathbf{r}_G' - \mathbf{r}_G)$ inside the integral over $\mathbf{r}_G'$. This leads to the intensity in the Gaussian image plane z_I as

$$
I_I(\mathbf{r}_I) = \int d^2\mathbf{r}_G\, I_G(\mathbf{r}_G)\, |H(\mathbf{r}_I - \mathbf{r}_G)|^2.
\tag{3.306}
$$

We define a new function

$$
J(\mathbf{r}_I - \mathbf{r}_G) = |H(\mathbf{r}_I - \mathbf{r}_G)|^2.
\tag{3.307}
$$

This leads to

$$
I_I(\mathbf{r}_I) = \int d^2\mathbf{r}_G\, I_G(\mathbf{r}_G)\, J(\mathbf{r}_I - \mathbf{r}_G).
\tag{3.308}
$$

Evidently, $J(\mathbf{r}_I - \mathbf{r}_G)$ represents the *intensity* point spread function for the special case of incoherent illumination.

It is useful to study this in the Fourier space of spatial frequencies. We define the Fourier transforms

$$\tilde{I}_I(\mathbf{K}) = \int d^2\xi \, I_I(\xi) \, e^{-i\mathbf{K}\cdot\xi}$$

$$\tilde{I}_G(\mathbf{K}) = \int d^2\xi \, I_G(\xi) \, e^{-i\mathbf{K}\cdot\xi}$$

$$\tilde{J}(\mathbf{K}) = \int d^2\xi \, J(\xi) \, e^{-i\mathbf{K}\cdot\xi}, \tag{3.309}$$

where $\mathbf{K}$ again represents 2π times the spatial frequency. Applying the convolution theorem (see Appendix A) to (3.308), we obtain

$$\tilde{I}_I(\mathbf{K}) = \tilde{I}_G(\mathbf{K}) \, \tilde{J}(\mathbf{K}). \tag{3.310}$$

In words, the transform of the image intensity is the transform of the ideal geometric image intensity, modulated by the transform of the point spread function. We now form the ratio

$$O(K) = \frac{\tilde{J}(\mathbf{K})}{\tilde{J}(0)}, \tag{3.311}$$

called the *optical transfer function* or OTF. Physically, it represents the normalized spatial frequency response of the optical system with respect to intensity. Its modulus $|O(K)|$ is called the *modulation transfer function* or MTF. We can gain an appreciation of the physical significance by relating $\tilde{J}(\mathbf{K})$ back to the amplitude transfer function (ATF) derived in the previous section. This was denoted $\tilde{H}(\mathbf{K})$. From (3.307, 3.309), we write

$$\tilde{J}(\mathbf{K}) = \int d^2\mathbf{r} \, |H(\mathbf{r})|^2 \, e^{-i\mathbf{K}\cdot\mathbf{r}}. \tag{3.312}$$

The amplitude transfer function $H(\mathbf{r})$ can be expressed in terms of its inverse Fourier transform as

$$H(\mathbf{r}) = \frac{1}{(2\pi)^2} \int d^2\mathbf{K} \, \tilde{H}(\mathbf{K}) \, e^{i\mathbf{K}\cdot\mathbf{r}}. \tag{3.313}$$

Substituting this into (3.312) and interchanging the order of integrations, we obtain

$$\tilde{J}(\mathbf{K}) = \frac{1}{(2\pi)^2} \int d^2\mathbf{K}' \, \tilde{H}(\mathbf{K}') \int d^2\mathbf{K}'' \, \tilde{H}^*(\mathbf{K}'')$$

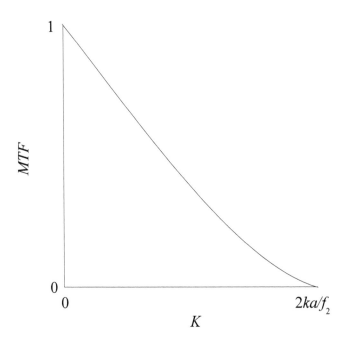

Figure 3.15: Modulation transfer function, round aperture, paraxial approximation.

$$\cdot \left[\frac{1}{(2\pi)^2} \int d^2 \mathbf{r} \, e^{-i(\mathbf{K}'' - \mathbf{K}' + \mathbf{K})} \right]. \tag{3.314}$$

We recognize the expression in square brackets as a Dirac delta function $\delta(\mathbf{K}'' - \mathbf{K}' + \mathbf{K})$, in which case

$$\tilde{J}(\mathbf{K}) = \frac{1}{(2\pi)^2} \int d^2 \mathbf{K}' \, \tilde{H}(\mathbf{K}') \, \tilde{H}^*(\mathbf{K}' - \mathbf{K}). \tag{3.315}$$

Mathematically, $\tilde{J}(\mathbf{K})$ is proportional to the autocorrelation function of $\tilde{H}(\mathbf{K})$ in $\mathbf{K}$-space. Substituting (3.297), we obtain

$$\tilde{J}(\mathbf{K}) = \frac{1}{(2\pi)^2} \int d^2 \mathbf{K}' \, P\left[-\frac{f_2}{k} \mathbf{K}' \right] P\left[-\frac{f_2}{k} (\mathbf{K}' - \mathbf{K}) \right], \tag{3.316}$$

recalling that M is the magnification, P is the pupil function, f_2 is the focal length of the final lens of the equivalent confocal system, and $k = 2\pi/\lambda$ is the wave number of the particle. We define a two-vector position $\mathbf{r}$ by

$$\mathbf{r} = -\frac{f_2}{k}\,\mathbf{K}, \qquad d^2\mathbf{K} = \left(\frac{k}{f_2}\right)^2 d^2\mathbf{r}. \qquad (3.317)$$

It follows that

$$\tilde{J}(\mathbf{K}) = \left(\frac{k}{2\pi f_2}\right)^2 \int d^2\mathbf{r}'\, P(\mathbf{r}')\, P(\mathbf{r}' - \mathbf{r}). \qquad (3.318)$$

Geometrically, the integral on the right is the common area of the pupil with itself displaced by $\mathbf{r}$.

An important special case is a round aperture, for which the pupil function is

$$P(\mathbf{r}) = 1 \qquad (3.319)$$

for $0 \le r \le a$, and zero for $r > a$. It follows that

$$\tilde{J}(\mathbf{K}) = \left(\frac{k}{2\pi f_2}\right)^2 \cdot 2a^2 \left[\cos^{-1}\left(\frac{r}{2a}\right) - \frac{r}{2a}\sqrt{1 - \left(\frac{r}{2a}\right)^2} \right]. \qquad (3.320)$$

This depends only on the magnitude $r = |\mathbf{r}|$, because of the radial symmetry of the pupil function $P(r)$. Substituting $r = f_2 K/k$, we obtain

$$\tilde{J}(K) = 2\left(\frac{ka}{2\pi f_2}\right)^2 \cdot \left[\cos^{-1}\left(\frac{f_2 K}{2ka}\right) - \frac{f_2 K}{2ka}\sqrt{1 - \left(\frac{f_2 K}{2ka}\right)^2} \right], \qquad (3.321)$$

where this is a function of the magnitude K. The optical transfer function is

$$O(K) = \frac{2}{\pi}\left[\cos^{-1}\left(\frac{f_2 K}{2ka}\right) - \frac{f_2 K}{2ka}\sqrt{1 - \left(\frac{f_2 K}{2ka}\right)^2} \right]. \qquad (3.322)$$

This is unity at $K = 0$ as required, and zero for

$$K_c = \frac{2ka}{f_2}, \qquad (3.323)$$

where K_c represents the cutoff value of K. No spatial frequencies above this value are transmitted by the optical system. We notice that a/f_2 is the tangent of the semiangle subtended by the pupil at the Gaussian image plane. The modulation transfer function (MTF) is plotted in Figure 3.15.

3.3.7   The wave front aberration function

An image point in the paraxial approximation is formed by a spherical wave converging on an ideal point in the Gaussian image plane. With aberrations present, the image is blurred and displaced from its ideal position. The aberrated image is formed by a wave which is distorted from an ideal spherical wave.

This is shown schematically in Figure 3.16. An ideal spherical wave front S_i fills the angular acceptance cone of the aperture. A ray emanates from every point along the wave front in a direction locally perpendicular to the wave front. These rays converge to an ideal point in the Gaussian image plane, as shown by the broken lines in the figure. The aberrated wave front S_a is locally displaced from the ideal wave front by a distance χ, which we designate the *wave front aberration function*. The rays emanting from the aberrated wave front converge to a region which is blurred and displaced from the ideal image point in general. These rays are depicted by the solid lines in the figure.

As discussed earlier, every optical system, however complicated, can be analyzed in terms of an equivalent system consisting of two ideal lenses. This is shown schematically in Figure 3.17. A point object is located in an object plane at O. The object plane coincides with the front focal plane of the first lens L_1. A physical aperture is located at the back focal plane of the first lens L_1, which coincides with the front focal plane of the second lens L_2.

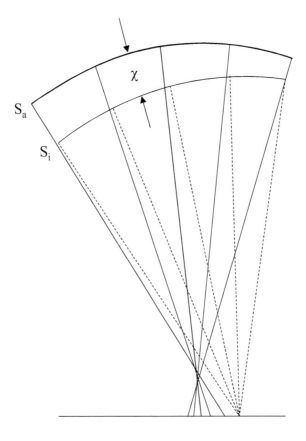

Figure 3.16: Wave front aberration.

The aperture plane is labeled A in the figure. In the absence of aberrations and in the limit of geometrical optics, an ideal point image I is formed from a point object O. The magnification M is defined as the ratio of the respective image and object heights. It is easy to verify by similar triangles in the figure that this is identical with the ratio $-f_2/f_1$ of the respective lens focal lengths, where the minus sign accounts for the inversion of the image.

Each ray emanating from the object O intersects the aperture plane A at a unique transverse position (x_A, y_A). Each point (x_A, y_A) in turn maps to a unique ray slope (x'_I, y'_I) where the ray intersects the Gaussian image plane. The wave front aberration

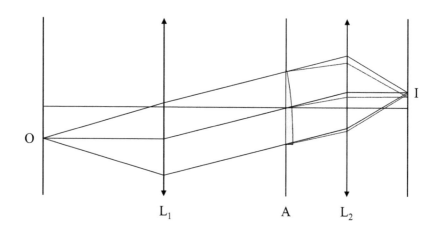

Figure 3.17: Equivalent confocal system.

function χ can be regarded as a function of transverse coordinates (x_A, y_A) in the aperture plane A.

The product $k\chi$ is the phase shift associated with the aberration, where k is the wave number given by $k = 2\pi/\lambda$. For a well-designed optical system χ is a small fraction of the wavelength λ. Equivalently, the phase shift due to aberrations is much less than 2π. All information about the aberrations is contained in the wave front aberration function χ. This is discussed in many books [11], [16], [67].

In the equivalent confocal system we consider the aberration to be entirely introduced in the aperture plane as an abrupt phase shift. This is conceptually equivalent to introducing a thin phase plate in the aperture plane, which shifts the phase by an amount $k\chi$. This is depicted in Figure 3.17.

The wave front aberration $\chi(x_A, y_A)$ is a scalar function defined in a plane. Assuming a round aperture, the function χ can in principle be expanded in a series of *Zernike polynomials*. Zernike polynomials are orthonormal functions defined on the unit disk $0 \le \rho \le 1$, where (ρ, ϕ) are polar coordinates in a plane. Here ρ is defined as the radial ray position in the aperture plane divided by the aperture radius in the equivalent confocal system. Any arbitrary function $F(\rho, \phi)$ which is defined on the unit disk $0 \leq \rho \leq 1$ and $0 \leq \phi \leq 2\pi$ can in principle be expanded as a linear combination of Zernike polynomials. The method is discussed below.

The Zernike polynomials $Z_n^m(\rho, \phi)$ are defined [1] by

$$
\begin{aligned}
Z_n^m(\rho, \phi) &= R_n^m(\rho) \cos(m\phi) \\
Z_n^{-m}(\rho, \phi) &= R_n^m(\rho) \sin(m\phi),
\end{aligned} \tag{3.324}
$$

where n and m are non-negative integers with $n \geq m$. The radial functions $R_n^m(\rho)$ are defined as

$$
R_n^m(\rho) = \sum_{k=0}^{(n-m)/2} \frac{(-1)^k (n-k)! \, \rho^{n-2k}}{k! \, [(n+m)/2 - k]! \, [(n-m)/2 - k]!} \tag{3.325}
$$

for $(n-m)$ even, and $R_n^m = 0$ for $(n-m)$ odd. It is easy to show that $R_n^m(1) = 1$, and therefore, $-1 \leq Z_n^m(\rho, \phi) \leq 1$.

The following orthogonality relations can be shown:

$$
\int_0^1 R_n^m(\rho) \, R_{n'}^m(\rho) \, \rho \, d\rho = \frac{1}{\sqrt{(2n+2)(2n'+2)}} \, \delta_{n,n'}
$$

$$
\int_0^{2\pi} \cos(m\phi) \cos(m'\phi) \, d\phi = \epsilon_m \, \pi \, \delta_{|m|,|m'|}
$$

$$
\int_0^{2\pi} \sin(m\phi) \sin(m'\phi) \, d\phi = (-1)^{m+m'} \, \pi \, \delta_{|m|,|m'|}, \quad (m \neq 0)
$$

$$
\int_0^{2\pi} \cos(m\phi) \sin(m'\phi) \, d\phi = 0, \tag{3.326}
$$

where $\delta_{i,j}$ is the Kronecker delta, and $\epsilon_m = 2$ if $m = 0$, and $\epsilon_m = 1$ if $m \neq 0$. It follows that

$$
\int_0^1 d\rho \, \rho \int_0^{2\pi} d\phi \, Z_n^m(\rho, \phi) \, Z_{n'}^{m'}(\rho, \phi) = \frac{\epsilon_m \, \pi}{2n+2} \, \delta_{n,n'} \, \delta_{m,m'}, \tag{3.327}
$$

where $(n-m)$ and $(n'-m')$ must both be even.

We define an arbitrary functon $F(\rho, \phi)$ as the linear combination of Zernike polynomials as follows:

$$F(\rho, \phi) = \sum_{n=0}^{\infty} \sum_{m=0}^{\infty} [\, a_{mn}\, Z_n^m(\rho, \phi) + b_{mn}\, Z_n^{-m}(\rho, \phi)\,], \qquad (3.328)$$

where the coefficients a_{mn} and b_{mn} are considered arbitrary to this point. This defines a Zernike transform. Using the above orthogonality relations it is possible to invert these equations as follows:

$$
\begin{aligned}
a_{mn} &= \frac{2n+2}{\epsilon_m\,\pi} \int_0^1 d\rho\,\rho \int_0^{2\pi} d\phi\, F(\rho, \phi)\, Z_n^m(\rho, \phi) \\
b_{mn} &= \frac{2n+2}{\epsilon_m\,\pi} \int_0^1 d\rho\,\rho \int_0^{2\pi} d\phi\, F(\rho, \phi)\, Z_n^{-m}(\rho, \phi). \quad (3.329)
\end{aligned}
$$

These two equations define the inverse Zernike transform.

The radial functions $R_n^m(\rho)$ are easily obtained by direct substitution of the various integer values n and m. The first nine are

$$
\begin{aligned}
R_0^0 &= 1 \\
R_1^1 &= \rho \\
R_2^0 &= 2\rho^2 - 1 \\
R_2^2 &= \rho^2 \\
R_3^1 &= 3\rho^3 - 2\rho \\
R_3^3 &= \rho^3 \\
R_4^0 &= 6\rho^4 - 6\rho^2 + 1 \\
R_4^2 &= 4\rho^4 - 3\rho^2 \\
R_4^4 &= \rho^4. \qquad (3.330)
\end{aligned}
$$

Substituting, the first fifteen Zernike polynomials with corresponding optical aberrations are as follows:

$$
\begin{aligned}
Z_0^0 &= 1 && \text{piston} \\
Z_1^{-1} &= \rho \sin \phi && y\text{-tilt} \\
Z_1^1 &= \rho \cos \phi && x\text{-tilt} \\
Z_2^{-2} &= \rho^2 \sin (2\phi) && \text{astigmatism}
\end{aligned}
$$

$$
\begin{aligned}
Z_2^0 &= 2\rho^2 - 1 & \text{defocus} \\
Z_2^2 &= \rho^2 \cos(2\phi) & \text{astigmatism} \\
Z_3^{-3} &= \rho^3 \sin(3\phi) & \text{trefoil} \\
Z_3^{-1} &= \rho(3\rho^2 - 2)\sin\phi & \text{coma} \\
Z_3^1 &= \rho(3\rho^2 - 2)\cos\phi & \text{coma} \\
Z_3^3 &= \rho^3 \cos(3\phi) & \text{trefoil} \\
Z_4^{-4} &= \rho^4 \sin(4\phi) \\
Z_4^{-2} &= \rho^2(4\rho^2 - 3)\sin(2\phi) \\
Z_4^0 &= 6\rho^4 - 6\rho^2 + 1 & \text{spherical} \\
Z_4^2 &= \rho^2(4\rho^2 - 3)\cos(2\phi) \\
Z_4^4 &= \rho^4 \cos(4\phi), & (3.331)
\end{aligned}
$$

where we recall that $0 \le \rho \le 1$ and $0 \le \phi \le 2\pi$.

It is useful in some cases to express these in Cartesian coordinates, again confined to the unit disk. To do this we first expand the trigonometric functions according to the well-known relations as follows:

$$
\begin{aligned}
\sin(2\phi) &= 2\sin\phi\cos\phi \\
\cos(2\phi) &= \cos^2\phi - \sin^2\phi \\
\sin(3\phi) &= \sin\phi(3\cos^2\phi - \sin^2\phi) \\
\cos(3\phi) &= \cos\phi(\cos^2\phi - 3\sin^2\phi) \\
\sin(4\phi) &= 4\sin\phi\cos\phi(\cos^2\phi - \sin^2\phi) \\
\cos(4\phi) &= \cos^4\phi - 6\sin^2\phi\cos^2\phi + \sin^4\phi. \quad (3.332)
\end{aligned}
$$

Substituting the Cartesian coordinates $x = \rho\cos\phi$ and $y = \rho\sin\phi$ we immediately obtain

$$
\begin{aligned}
Z_0^0 &= 1 \\
Z_1^{-1} &= y \\
Z_1^1 &= x \\
Z_2^{-2} &= 2xy \\
Z_2^0 &= 2(x^2 + y^2) - 1
\end{aligned}
$$

$$
\begin{aligned}
Z_2^2 &= x^2 - y^2 \\
Z_3^{-3} &= y\,(3\,x^2 - y^2) \\
Z_3^{-1} &= y\,[\,3\,(x^2 + y^2) - 2\,] \\
Z_3^1 &= x\,[\,3\,(x^2 + y^2) - 2\,] \\
Z_3^3 &= x\,(x^2 - 3\,y^2) \\
Z_4^{-4} &= 4\,x\,y\,(x^2 + y^2) \\
Z_4^{-2} &= 2\,x\,y\,[\,4\,(x^2 + y^2) - 3\,] \\
Z_4^0 &= 6\,(x^2 + y^2)^2 - 6\,(x^2 + y^2) + 1 \\
Z_4^2 &= (x^2 - y^2)\,[\,4\,(x^2 + y^2) - 3\,] \\
Z_4^4 &= x^4 - 6\,x^2 y^2 + y^4.
\end{aligned}
\tag{3.333}
$$

Our goal is to express an arbitrary phase shift $k\chi(x_A, y_A)$ in terms of a corresponding set of Zernike coefficients (a_{mn}, b_{mn}), where $k = 2\pi/\lambda$ is the wave number. To this end we define a set of coordinate transformations as follows:

$$
\begin{aligned}
x &= \rho \cos\phi = x_A/R_A \\
y &= \rho \sin\phi = y_A/R_A \\
\rho &= \sqrt{x^2 + y^2},
\end{aligned}
\tag{3.334}
$$

where R_A is the radius of the aperture in the equivalent confocal system. Based on this, we define the formal functional substitutions

$$
\begin{aligned}
G(x, y) &= F(\rho, \phi) \\
k\chi(x_A, y_A) &= G(x, y).
\end{aligned}
\tag{3.335}
$$

Finally, we define the correspondence between the physical system and the equivalent confocal system according to

$$
\begin{aligned}
x_I' &= x_A/f_2 \\
y_I' &= y_A/f_2 \\
M &= -f_2/f_1,
\end{aligned}
\tag{3.336}
$$

where we assume the final ray slopes (x'_I, y'_I) and the magnification M are identical for the physical system and the equivalent confocal system.

This completes the correspondence between the aberrations of the physical system and the Zernike coefficients (a_{mn}, b_{mn}). The Zernike coefficients uniquely specify the aberrations. Strictly speaking, these coefficients are a function of object position O. In a well-designed system, the coefficients do not vary greatly across the object plane.

3.3.8 Relationship between diffraction and the Heisenberg uncertainty principle

We have seen in the foregoing sections that diffraction and interference follow naturally from the Fresnel–Kirchhoff relation (3.244). In turn, this relation represents a stationary-state solution of the spatial part of Schrödinger's equation (3.230) for a free particle wave function. It is of great interest to consider diffraction and interference from a closely related point of view, namely, Heisenberg's uncertainty principle. This is the subject of the present section.

We begin with diffraction from two parallel slits. This is shown schematically in Figure 3.18. A plane wave is incident from the top of the figure, with propagation direction normally incident on a screen S. The screen has two infinitely long parallel slits, oriented out of the page. A diffracted ray I emanates from the left slit, and a diffracted ray II emanates from the right slit. The two rays are assumed to be parallel to one another, and propagate at an angle θ relative to the central axis. A thin lens L causes the two rays to converge at a viewing plane P. The axial spacing between S and L

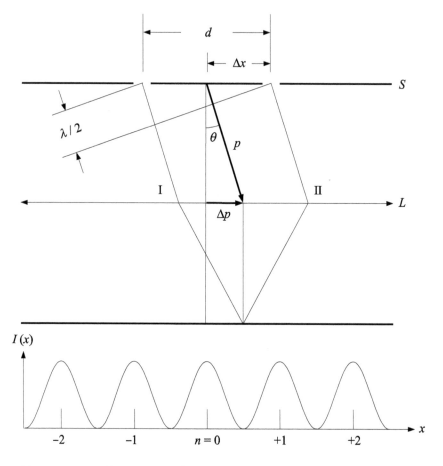

Figure 3.18: Two-slit diffraction with destructive interference.

is assumed to be equal to the axial spacing between L and P, and both are equal to the focal length of the lens. The plane P thus represents the diffraction or Fourier plane of the lens. In this configuration every viewing angle θ is mapped to a unique transverse coordinate x in the plane P.

Scanning the viewing angle θ gives rise to an intensity distribution $I(x)$ in the plane P. The functional form for $I(x)$ follows directly from (3.279). It is left as an exercise to solve for $I(x)$. The peaks represent constructive interference of the two waves, and the valleys represent destructive interference. It is evident from the figure that the two rays have a path length difference given by $d \sin \theta$, where d is the separation distance of the two slits. Constructive interference occurs when this path length difference is equal to an integral number of wavelengths λ. Equivalently,

$$d \sin \theta = n \lambda, \qquad n = 0, \pm 1, \pm 2, \pm 3, \ldots . \qquad (3.337)$$

Destructive interference occurs when the path length difference is equal to a half-odd number of wavelengths. Equivalently,

$$d \sin \theta = \left(n + \tfrac{1}{2}\right) \lambda, \qquad n = 0, \pm 1, \pm 2, \pm 3, \ldots . \qquad (3.338)$$

Next we consider the specific viewing angle θ for which the intensity distribution $I(x)$ has its first minimum. This is shown in the figure, where the path length difference between the two rays is $\lambda/2$, and $n = +1$ in (3.338). We assume the particle has momentum p, which is related to its de Broglie wavelength λ by

$$p = \frac{h}{\lambda}, \qquad (3.339)$$

where h is Planck's constant. This momentum has a transverse component Δp, which satisfies

$$\frac{\Delta p}{p} = \sin \theta. \qquad (3.340)$$

Since Δp represents the half-width of the first interference fringe with $n = 0$, we ascribe an uncertainty Δp to the transverse momentum of the particle.

Separately, one has no knowledge about which of the two slits the particle passed through. Following Feynman [30, Chapter 1, Volume 3], any attempt to measure which slit the particle passed through would perturb the wave function, thereby irreparably destroying the interference. Consequently, we ascribe an uncertainty $\Delta x = d/2$ to the transverse position of the particle. It is left as a brief exercise to show (3.338, 3.339, 3.340) that

$$\Delta x \, \Delta p = \frac{h}{4}. \qquad (3.341)$$

This represents a statement of Heisenberg's uncertainty principle.

The assignments of Δx and Δp are somewhat arbitrary. For example, one might assign the full-widths, instead of the half-widths for Δx and Δp, in which case the right-hand side in (3.340) would be multiplied by a factor of four. The uncertainty relation (3.341) is therefore properly regarded as an approximation.

The relation (3.339) applies to a photon, and equivalently to a massive particle. It follows that the uncertainty relation (3.341) applies to both a photon and a massive particle. Alternatively, the momentum can be expressed as

$$p = \hbar k \qquad (3.342)$$

for a photon and massive particle, where k is the wave number given as $k = 2\pi/\lambda$. The uncertainty relation (3.341) becomes

$$\Delta x \, \Delta k \approx \frac{\pi}{2}, \qquad (3.343)$$

given the above assumptions. This can be regarded as a general property of waves, not just a photon or particle. It applies to many wavelike phenomena, including water waves and sound waves, as examples. The usual coherent superposition of waves with different values of k to form a wave packet, applies to these different types of waves as well.

Chapter 4

Particle scattering

The interaction of fast charged particles with matter provides the basic physical mechanism underlying most applications of charged particle beam instruments. In this interaction, an incident particle strikes a target, transfering momentum and energy. The incident particle is completely characterized by its rest mass, charge, momentum, and spin polarization. The target can consist of bulk material (solid, liquid, or vapor), a single atom, a molecule, or a second particle (composite or pointlike). The interaction can be governed by the strong, weak, Coulomb, or gravitational forces. The gravitational force is too weak to be important for charged particles on the laboratory scale of dimensions. However the classical Kepler problem is formally identical to Coulomb scattering between two individual charged particles via the inverse square dependence of the instantaneous force on the separation. In all cases, the interaction can be used to probe the physical or chemical properties of the target.

In some cases where the rest mass of the incident particle is much smaller than the target particle, the incident particle transfers negligible energy to the target. Such an event is known as *elastic* scattering. An example is the angular deflection of a fast electron by the screened Coulomb potential of an atomic nucleus. This provides the basic contrast mechanism of a transmission electron microscope. In elastic scattering, the phase relationship of the incident particle before and after the scattering event is preserved. This phase coherence enables the study of a crystalline target through electron diffraction.

A separate example of nearly elastic scattering is the Coulomb scattering of an energetic alpha particle by a heavy nucleus. This is the mechanism originally used by Rutherford [59] to characterize an atomic nucleus as a compact, massive, multiply charged body. In elastic scattering, the magnitude of the momentum of the incident particle is unchanged by the scattering, but the direction can be significantly changed. Consequently, significant momentum can be transferred.

Alternatively, the incident particle can transfer significant energy to the target. Such an event is known as *inelastic* scattering. This is much more complex than elastic scattering, because a rich variety of secondary processes can result. For example, an incident fast electron can excite the electronic states of a target material, giving rise to an excited-state atom, secondary photon, free electron, electron-hole pair (exciton), or plasmon. Performing electron energy loss spectroscopy on the scattered electron yields information on the chemical and physical nature of the target material.

Separately, the resulting secondary electron current forms the basic contrast mechanism in the scanning electron microscope, or a scanning helium ion microscope. Alternatively, bombarding a material with an ion beam causes secondary ions of the target material to be ejected. Performing secondary ion mass spectrometry gives direct information about the atomic composition of the target material.

Separately still, an incident electron or ion beam can be used to chemically alter a target polymer film in a useful way. The patterned film is then used as a binary mask in the process known as *lithography*. A focused electron beam or a focused helium ion beam can form a very fine, sharp writing pencil. Lithographic structures down to a few nanometers in size have been produced by this method. Ultimate lithographic resolution is limited by the interaction of the incident particle with the recording medium. Improving this resolution remains a topic of current research.

Alternatively, a focused ion beam can selectively remove material from a bulk sample, enabling fabrication of useful structures. A focused electron beam can be used to simultaneously in the same (dual beam) instrument to produce an SEM image, thus enabling in situ observation and endpoint detection of this removal process.

Yet another important example of inelastic scattering is the production of a host of particle species in a high energy particle accelerator. This process takes place by the conversion of kinetic energy of the incident particle to mass of the products. This has been used in ongoing research to deduce the most fundamental properties of elementary particles and their interactions.

An enormous literature exists describing the interaction of charged particles with matter. Rather than attempt a comprehensive summary, we will confine our attention to two-particle scattering. This is the most basic of all scattering processes, and is derived from first principles of physics in a straightforward way. Two-particle scattering forms the basis of many of the interactions of charged particle beams with matter in practical instruments.

The central problem in the following sections is to calculate the momentum and energy transfer resulting from two-particle scattering. Closely related to this is the intensity as a function of scattering angle measured relative to the direction of the incident particle, where this represents the relevant measurable quantity. Although the process is fundamentally quantum mechanical, a great deal of intuitive understanding can be gained by first studying the classical description, and then proceeding to the quantum mechanical description.

4.1   Classical particle kinematics

The generic two-particle scattering problem assumes a single particle with vector kinetic momentum $\mathbf{p}$, incident on a second particle
which is initially at rest in the lab frame. The first particle transfers
momentum and energy to the second particle, and both particles
exit to a final state. The final state is governed by conservation of
total momentum and energy, as well as the details of the interaction. The first particle is assumed to come from a distance which
is large compared with the dimensions of the interaction volume
of the two particles. Similarly, both particles exit to a large distance in the final state. At large distances the potential energy of
the interaction can be ignored. In this section we investigate the
constraints imposed by momentum and energy conservation. This
general topic is referred to as *kinematics*.

We assume the incident particle has rest mass m_1, vector kinetic
momentum $\mathbf{p}_1$, and total energy E_1. We assume the stationary
particle has rest mass m_2, vector kinetic momentum $\mathbf{p}_2$, and total
energy E_2. These quantities are related by

$$\begin{aligned} E_1^2 &= p_1^2 c^2 + m_1^2 c^4 \\ E_2^2 &= p_2^2 c^2 + m_2^2 c^4, \end{aligned} \tag{4.1}$$

consistent with special relativity. We assign the value $\mathbf{p}_1 = \mathbf{p}$ for
the incident particle, where $\mathbf{p}$ is assumed to be known *a priori*,
along with the rest masses m_1 and m_2. We assign the value $\mathbf{p}_2 = 0$
for the stationary target particle. We assume that neither particle has internal degrees of freedom. By implication, we ignore the
effects of spin in the following analysis. We further assume that
each particle retains its original rest mass through the collision.

Since the only force is that which acts between the two particles, it follows that the center of mass moves at constant velocity. The motion of the center of mass is therefore uninteresting.
We seek a frame of reference such that the total momentum of
the two-particle system is zero. We call this system the center-of-
momentum or CM frame. The scattering is most simply analyzed

in the CM frame, since the superfluous motion of the center of mass does not appear. This is shown in Figure 4.1, where the top diagram represents the lab frame, and the middle diagram represents the CM frame. We denote the CM frame by primed quan-

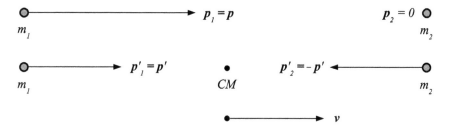

Figure 4.1: Reference frames for scattering.

tities, and the lab frame by unprimed quantities. We assign the value **v** to the vector velocity of the CM measured in the lab frame.

The incident momentum and energy are measured in the lab frame, the scattering probability is calculated in the CM frame, and the scattered intensity is measured as a function of scattering angle in the lab frame. Our procedure must therefore consist of transforming from the lab to the CM frame, then calculating the scattering probability as a function of scattering angle in the CM frame, then finally transforming back to the lab frame. We adopt the notation

$$
\begin{aligned}
\beta &= v/c \\
\gamma &= \frac{1}{\sqrt{1 - v^2/c^2}} = \frac{1}{\sqrt{1 - \beta^2}}.
\end{aligned}
$$

For brevity of notation, we further make the following substitutions for the mass and momentum, respectively:

$$
\begin{aligned}
m c^2 &\rightarrow m \\
pc &\rightarrow p.
\end{aligned} \tag{4.2}
$$

This is equivalent to a system of units where the speed of light is $c = 1$. The reader can transform back to the original quantities at any point in the calculation.

Lorentz transformation from lab (unprimed) to CM (primed) frame gives

$$
\begin{aligned}
p_1' &= \gamma\,(p_1 - \beta\,E_1) \\
E_1' &= \gamma\,(E_1 - \beta\,p_1) \\
p_2' &= \gamma\,(p_2 - \beta\,E_2) \\
E_2' &= \gamma\,(E_2 - \beta\,p_2)
\end{aligned}
\tag{4.3}
$$

for the initial state. This makes use of the fact that momentum and energy form a four-vector $p_\mu = (\mathbf{p}, iE/c)$. Substituting the assumed values $p_1 = p$, $p_2 = 0$, $E_2 = m_2$, $p_1' = p'$, and $p_2' = -p'$, we obtain

$$
\begin{aligned}
p' &= \gamma\,(p - \beta\,E_1) \\
E_1' &= \gamma\,(E_1 - \beta\,p) \\
-p' &= \gamma\,(-\beta\,m_2) \\
E_2' &= \gamma\,m_2.
\end{aligned}
\tag{4.4}
$$

Solving the first and third equations for β, we obtain

$$
\beta = \frac{p}{E_1 + m_2} = \frac{\sqrt{E_1^2 - m_1^2}}{E_1 + m_2}
\tag{4.5}
$$

and

$$
\gamma = \frac{E_1 + m_2}{\sqrt{m_1^2 + m_2^2 + 2E_1 m_2}}.
\tag{4.6}
$$

Substituting these into the second and fourth equations of (4.4) yields the energies of the two individual particles in the CM frame, given respectively by

$$
\begin{aligned}
E_1' &= \frac{m_1^2 + E_1 m_2}{\sqrt{m_1^2 + m_2^2 + 2E_1 m_2}} \\
E_2' &= \frac{m_2^2 + E_1 m_2}{\sqrt{m_1^2 + m_2^2 + 2E_1 m_2}}.
\end{aligned}
\tag{4.7}
$$

The sum E' of the energies of the two particles in the CM frame is thus given by

$$E' = E'_1 + E'_2 = \sqrt{m_1^2 + m_2^2 + 2E_1 m_2}. \tag{4.8}$$

We thus obtain several identities which will prove useful later:

$$\begin{aligned} E'_1 E' &= m_1^2 + E_1 m_2 \\ E'_2 E' &= m_2^2 + E_1 m_2 \\ p' E' &= m_2 p, \end{aligned} \tag{4.9}$$

where

$$E'^2 = (E_1 + m_2)^2 - p^2 = m_1^2 + m_2^2 + 2E_1 m_2. \tag{4.10}$$

We have thus succeeded in calculating all relevant initial quantities p', E'_1, and E'_2 in the CM frame from known quantities p, m_1, m_2, E_1 in the lab frame.

We notice that in the CM frame, the total kinetic energy T' is given in terms of the total energy E' by

$$T' = E' - (m_1 + m_2) = \left(m_1^2 + m_2^2 + 2\,E_1\,m_2\right)^{1/2} - (m_1 + m_2). \tag{4.11}$$

In words, the kinetic energy is the total energy minus the energy of the rest masses. Separately, the total energy E_1 of the incident particle in the lab frame can be written as

$$E_1 = \gamma_1\,m_1, \tag{4.12}$$

where $\gamma_1 = 1/\sqrt{1 - \beta_1^2}$ applies to the initial velocity $\beta_1 = v_1/c$ of the incident particle measured in the lab frame. This velocity v_1 is not to be confused with the velocity v of the CM measured in the lab frame. Substituting,

$$\begin{aligned} T' &= \left[(m_1 + m_2)^2 + 2\,(\gamma_1 - 1)\,m_1\,m_2\right]^{1/2} - (m_1 + m_2) \\ &= (m_1 + m_2)\left[1 + 2\,(\gamma_1 - 1)\,\frac{m_1\,m_2}{(m_1 + m_2)^2}\right]^{1/2} - (m_1 + m_2). \end{aligned} \tag{4.13}$$

In the low energy limit, $\beta_1 \ll 1$. Keeping only the lowest order terms in the Taylor series, this reduces to

$$T' = \tfrac{1}{2}\left(\frac{m_1 m_2}{m_1 + m_2}\right)\beta_1^2. \qquad (4.14)$$

We now define a quantity called the *reduced mass* M, given by

$$M = \frac{m_1 m_2}{m_1 + m_2}, \qquad (4.15)$$

from which we write

$$T' = \tfrac{1}{2} M \beta_1^2. \qquad (4.16)$$

We conclude from this that the two-body scattering problem in the CM frame is mathematically equivalent to a single particle of mass M scattering about a fixed point, which coincides with the center of momentum. This reduction of the scattering problem is called the *equivalent one-body problem.*

We are now in a position to consider the final state after scattering. This is shown in the CM frame in Figure 4.2, where the particle with rest mass m_1 has final momentum $\mathbf{q}_1'$, and the particle with rest mass m_2 has final momentum $\mathbf{q}_2'$. We assume total energies ε_1' and ε_2' for the two particles after scattering, where

$$\begin{aligned} \varepsilon_1'^{\,2} &= q_1'^{\,2} + m_1^2 \\ \varepsilon_2'^{\,2} &= q_2'^{\,2} + m_2^2. \end{aligned} \qquad (4.17)$$

The scattering angle in the CM frame is defined as θ'. In later sections, we will address the central scattering problem, namely, calculation of the scattered intensity as a function of θ'. Consequently, we assume θ' to be known for now from the calculation to come.

Since measurement is always performed in the lab frame, we must express the relevant quantities there. This is shown in Figure 4.3, where the incident particle with mass m_1 is assumed to scatter through angle θ_1, and the incident particle with mass m_2 is assumed to scatter through angle θ_2. We wish to calculate the scat-

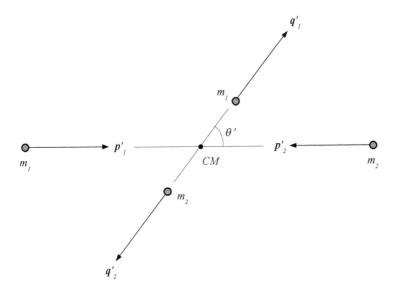

Figure 4.2: Initial and final states, CM frame.

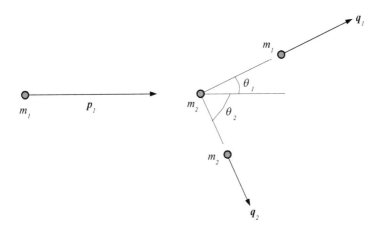

Figure 4.3: Initial and final states, lab frame.

tering angles θ_1 and θ_2 in the lab frame in terms of the scattering angle θ' in the CM frame. This is accomplished in principle by Lorentz transformation from the CM to lab frame as follows:

$$q_{1z} = \gamma\left(q'_{1z} + \beta\,\varepsilon'_1\right)$$

$$\varepsilon_1 = \gamma(\varepsilon_1' + \beta q_{1z}')$$
$$q_{2z} = \gamma(q_{2z}' + \beta \varepsilon_2')$$
$$\varepsilon_2 = \gamma(\varepsilon_2' + \beta q_{2z}'), \tag{4.18}$$

where we have replaced β by $-\beta$, and interchanged primed and un-primed quantities in the earlier Lorentz transformation. We have assumed that the relative velocity of the two reference frames is along the z-axis. Thus only the z-components of the momenta are altered by the Lorentz transformation, while the transverse components remain unaltered.

We *assume* that, in the CM frame, the scattering can be described by

$$q_1' = p_1', \qquad \varepsilon_1' = E_1' \tag{4.19}$$

for the particle with rest mass m_1, and

$$q_2' = p_2', \qquad \varepsilon_2' = E_2' \tag{4.20}$$

for the particle with rest mass m_2. These conditions express conservation of the magnitude of momentum for each particle *individually*. By implication, total energy is conserved for each particle individually. This is valid in the CM frame, but not in the lab frame. Substituting above, we obtain

$$q_{1z} = \gamma[p'\cos\theta' + \beta E_1']$$
$$\varepsilon_1 = \gamma[E_1' + \beta p'\cos\theta']$$
$$q_{2z} = \gamma[p'\cos(\pi - \theta') + \beta E_2']$$
$$\varepsilon_2 = \gamma[E_2' + \beta p'\cos(\pi - \theta')]. \tag{4.21}$$

The transverse x-components of the final momentum are transformed as follows:

$$q_{1x} = q_{1x}' = p'\sin\theta'$$
$$q_{2x} = q_{2x}' = -p'\sin\theta'. \tag{4.22}$$

From this we obtain

$$\tan\theta_1 = \frac{q_{1x}}{q_{1z}} = \frac{p'\sin\theta'}{\gamma(p'\cos\theta' + \beta E_1')}$$
$$\tan\theta_2 = \frac{q_{2x}}{q_{2z}} = \frac{-p'\sin\theta'}{\gamma(-p'\cos\theta' + \beta E_2')}. \tag{4.23}$$

We now make use of

$$\frac{\beta\,E_1'}{p'} = \frac{m_1^2 + E_1\,m_2}{m_2^2 + E_1\,m_2} \equiv \alpha$$

$$\frac{\beta\,E_2'}{p'} = \frac{m_2^2 + E_1\,m_2}{m_2^2 + E_1\,m_2} = 1. \tag{4.24}$$

We have defined a new dimensionless quantity α in terms of quantities which are all known. Substituting, we express the final scattering angles θ_1 and θ_2 in the lab frame, in terms of the known CM scattering angle θ' as

$$\tan\theta_1 = \frac{(m_1^2 + m_2^2 + 2\,E_1\,m_2)^{1/2}\,\sin\theta'}{(E_1 + m_2)\,(\cos\theta' + \alpha)}$$

$$\tan\theta_2 = \frac{(m_1^2 + m_2^2 + 2\,E_1\,m_2)^{1/2}\,\sin\theta'}{(E_1 + m_2)\,(\cos\theta' - 1)}. \tag{4.25}$$

These two equations express the scattering angles in the lab frame in terms of quantities which are all known. This represents the main result to this point.

It is of great interest to investigate the momentum and energy transferred in the lab frame. These are found by subtracting the initial state values from the final state values. This is embodied in the equations

$$\Delta\mathbf{p}_1 = \mathbf{q}_1 - \mathbf{p}_1$$
$$\Delta\mathbf{p}_2 = \mathbf{q}_2 - \mathbf{p}_2$$

$$\Delta E_1 = \varepsilon_1 - E_1$$
$$\Delta E_2 = \varepsilon_2 - E_2, \tag{4.26}$$

where the subscripts 1 and 2 refer to the incident and target particles, respectively, and where $\Delta\mathbf{p}_i$ is the transferred vector kinetic momentum and ΔE is the transferred total energy. Since the scattering takes place in a single plane, the momentum $\mathbf{p}$ is a two-vector. In the following, we label the direction of the incident particle momentum as the z-axis, and the orthogonal axis as the

x-axis. It is straightforward, but quite tedious to calculate the momentum and energy transfer. It is left as an exercise to the reader to set up the algebra, based on the above analysis. Indeed, an ambitious reader could carry this through to a closed-form solution.

The problem becomes greatly simpler in the nonrelativistic approximation. This approximation is relevant to a large variety of charged particle instruments, which operate at low energy, where the kinetic energy is small relative to the particle rest-mass energy. This approximation also provides significant intuitive insight into the scattering process.

We continue to use the same notation for the initial state in the lab frame, namely, we assume that particle 1 (the incident particle) has rest mass m_1, vector kinetic momentum $\mathbf{p}_1$, and total energy E_1. We assume that particle 2 (the target particle) has rest mass m_2, vector kinetic momentum $\mathbf{p}_2$, and total energy E_2.

We also continue to use the same notation for the final state after scattering, namely, we assume that particle 1 has rest mass m_1, vector kinetic momentum $\mathbf{q}_1$, and total energy ε_1. We assume that particle 2 has rest mass m_2, vector kinetic momentum $\mathbf{q}_2$, and total energy ε_2.

In the nonrelativistic limit these quantities are related by

$$
\begin{aligned}
E_1 &= \frac{p_1^2}{2m_1} \\[2mm]
E_2 &= \frac{p_2^2}{2m_2} \\[2mm]
\varepsilon_1 &= \frac{q_1^2}{2m_1} \\[2mm]
\varepsilon_2 &= \frac{q_2^2}{2m_2}.
\end{aligned}
\tag{4.27}
$$

We assume the initial condition in the lab (unprimed) frame that $\mathbf{p}_1 = \mathbf{p}$ and $\mathbf{p}_2 = 0$, where $\mathbf{p}$ is oriented along the $+z$ axis, and is known *a priori*.

With these assumptions, we now proceed to find $\mathbf{q}_1$ and $\mathbf{q}_2$, making use of conservation of momentum and energy. As before, we transform from the lab to the CM frame, then calculate the scattering probability as a function of scattering angle in the CM frame, then finally transform back to the lab frame. By definition $p_2' = -p_1'$ in the (primed) CM frame, since by definition the total momentum is zero in this frame. Separately, the individual particle velocities in the CM frame are given in terms of the velocities in the lab frame by

$$
\begin{aligned}
v_1' &= v_1 - v \\
v_2' &= v_2 - v,
\end{aligned}
\tag{4.28}
$$

where v is the velocity of the CM, measured in the lab frame. From the above assumptions, it is left as an exercise to the reader to show that the relative velocity of the two frames is given by

$$
v = \frac{p}{m_1 + m_2}.
\tag{4.29}
$$

Further, it follows that

$$
\begin{aligned}
p_1' &= p\,\frac{m_2}{m_1 + m_2} \\
p_2' &= -p\,\frac{m_2}{m_1 + m_2},
\end{aligned}
\tag{4.30}
$$

which satisfies the condition that $p_2' = -p_1'$ as required for the CM frame.

Next we invoke the condition that the scalar kinetic momentum is preserved in the scattering for each particle individually in the CM frame, that is,

$$
\begin{aligned}
q_1' &= p_1' \\
q_2' &= p_2'.
\end{aligned}
\tag{4.31}
$$

This is consistent with the fact that the vector momenta $\mathbf{q}_1'$ and $\mathbf{q}_2'$ after the collision are equal and opposite in the CM frame.

Taking account of these, we resolve the momenta $\mathbf{q}'_1$ and $\mathbf{q}'_2$ for the respective particles into transverse x-components, and longitudinal z-components:

$$
\begin{aligned}
q'_{1x} &= p'_1 \sin \theta' \\
q'_{1z} &= p'_1 \cos \theta' \\
q'_{2x} &= -p'_1 \sin \theta' \\
q'_{2z} &= -p'_1 \cos \theta',
\end{aligned}
\tag{4.32}
$$

where $0 \le \theta' \le \pi$. Next, we transform to the lab frame. By velocity addition of the longitudinal components only,

$$
\begin{aligned}
q_{1x} &= q'_{1x} \\
q_{1z} &= q'_{1z} + m_1 v \\
q_{2x} &= q'_{2x} \\
q_{2z} &= q'_{2z} + m_2 v.
\end{aligned}
\tag{4.33}
$$

Substituting, it follows that

$$
\begin{aligned}
q_{1x} &= \frac{pm_2}{m_1 + m_2} \sin \theta' \\
q_{1z} &= \frac{pm_2}{m_1 + m_2} \left(\cos \theta' + \frac{m_1}{m_2} \right) \\
q_{2x} &= -\frac{pm_2}{m_1 + m_2} \sin \theta' \\
q_{2z} &= -\frac{pm_2}{m_1 + m_2} \left(\cos \theta' - 1 \right).
\end{aligned}
\tag{4.34}
$$

This represents the solution for the final momenta, where the right-hand sides consist of all known quantities.

It is straightforward to calculate the momentum transferred to the two particles in the lab frame, $\Delta \mathbf{p}_i \equiv \mathbf{q}_i - \mathbf{p}_i$. This is

$$
\begin{aligned}
\Delta p_{1x} &= \frac{pm_2}{m_1 + m_2} \sin \theta' \\
\Delta p_{1z} &= \frac{pm_2}{m_1 + m_2} (\cos \theta' - 1)
\end{aligned}
$$

$$\Delta p_{1x} = -\frac{pm_2}{m_1 + m_2} \sin \theta'$$

$$\Delta p_{1z} = -\frac{pm_2}{m_1 + m_2} (\cos \theta' - 1). \quad (4.35)$$

The energy transferred between the two particles in the lab frame, $\Delta E_i \equiv \varepsilon_i - E_i$ follows immediately as

$$\Delta E_1 = \frac{p^2 m_2}{(m_1 + m_2)^2} (\cos \theta' - 1)$$

$$\Delta E_2 = -\frac{p^2 m_2}{(m_1 + m_2)^2} (\cos \theta' - 1). \quad (4.36)$$

The scattering angles θ_1 and θ_2 in the lab frame are easily found to obey

$$\tan \theta_1 = \frac{\sin \theta'}{(\cos \theta' + m_1/m_2)}$$

$$\tan \theta_2 = \frac{\sin \theta'}{(\cos \theta' - 1)}. \quad (4.37)$$

This agrees with the earlier relativistic result in the limit where the kinetic energy is negligible compared with the rest mass. Mathematically, this is equivalent to $E_i \approx m_i$. The reader is encouraged to verify these results.

We have thus succeeded in calculating the momentum and energy transfer in closed form, in the nonrelativistic limit. We have also calculated the scattering angles θ_1 and θ_2 in the lab frame, in the relativistic case, and the nonrelativistic approximation. This represents the complete solution to the two-particle scattering kinematics. This will prove very useful in the following sections.

4.2 Scattering cross section and classical scattering

Having obtained the reduction to the equivalent one-body problem, we are now in a position to address our central problem, namely, calculation of the scattered intensity as a function of scattering angle. For simplicity of notation in the following, we make the substitution

$$\vartheta \equiv \theta' \qquad (4.38)$$

for the scattering angle in the CM frame. From the preceding analysis, this is equivalent to the scattering angle of the reduced mass M from the fixed scattering center in the CM frame.

It is instructive to first study elastic scattering in the context of classical mechanics. The geometry is shown schematically in Figure 4.4 for a repulsive scattering force. The particle of mass M

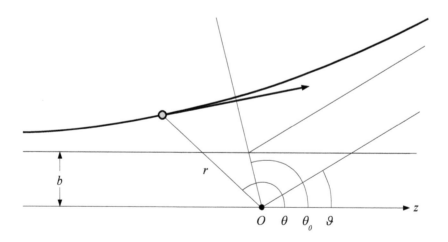

Figure 4.4: Classical elastic scattering geometry.

in the equivalent one-body problem is incident from the left, and initially travels along the $+z$ direction. The fixed scattering center at O causes the particle to trace out a curved trajectory given by $r(\theta)$. The particle passes to a very large distance, where the scattering force becomes negligible. The resulting scattering angle is ϑ, not to be confused with the instantaneous anglar coordinate θ.

We assume a uniformly dense beam of many particles incident on the scattering center from the left. We expect that the number of scattered particles dN detected at angle ϑ in a time interval dt must be proportional to the product of the incident intensity S_0 times the scattered solid angle element $d\Omega$ times the time interval dt, i.e.,

$$dN = \sigma(\vartheta)\, S_0\, d\Omega\, dt, \tag{4.39}$$

where $\sigma(\vartheta)$ is a proportionality factor which depends on the scattering angle ϑ. This factor contains all relevant information about the details of the scattering process. It is called the *differential cross section*. Rearranging factors, this is

$$\sigma(\vartheta) = \frac{1}{S_0} \frac{dN}{d\Omega\, dt}. \tag{4.40}$$

The differential cross section is the number of scattered particles per unit solid angle, per unit time, per unit incident intensity. It has units of area. Mathematically, the central problem is to find the differential cross section $\sigma(\vartheta)$.

The strength of the scattering and resulting ϑ vary inversely with the distance b, called the *impact parameter*. The problem is axially symmetric about the z-axis. Considering the range of possible values of b, we therefore write

$$S_0 = \frac{dN}{dA_0\, dt} = \frac{1}{2\pi b\, db} \frac{dN}{dt}, \tag{4.41}$$

where the cross-sectional area element dA_0 is an annulus centered on the z-axis. The final solid angle element $d\Omega$ is given by

$$d\Omega = \frac{dA}{r^2} = 2\pi\, \sin\vartheta\, d\vartheta, \tag{4.42}$$

where the area element dA is an annulus on a sphere of very large radius centered about the scattering center at O. Substituting (4.41, 4.42) into (4.40), the differential cross section is then

$$\sigma(\vartheta) = \frac{b}{\sin\vartheta} \frac{db}{d\vartheta}. \tag{4.43}$$

This applies to any case with axial symmetry, regardless of the detailed dependence of the scattering force on the separation r.

We assume that all relevant information about the scattering forces is contained in the potential energy $U(\mathbf{x})$ between the two particles, where $\mathbf{x}$ is the three-vector spatial separation between the particles. We assume U to be known. For the present analysis we now consider the special case for which the potential energy can be written as

$$U(r) = \frac{q_1 q_2}{4\pi\epsilon_0 r} \equiv \frac{\kappa}{r}, \tag{4.44}$$

where $r = |\mathbf{x}|$. The potential energy is inversely proportional to the magnitude of the separation r between the two particles, and is spherically symmetric. This is the electrostatic potential energy arising from the Coulomb interaction between two charges q_1 and q_2 separated by a distance r. With charges of opposite sign, $\kappa < 0$, giving rise to an attractive force. With charges of like sign, $\kappa > 0$, giving rise to a repulsive force. This is just the classical Kepler problem. An equation for the trajectory is expressed in polar coordinates as the radius r as a function of the scattering angle θ. This was derived previously in the section on applications of Hamilton–Jacobi theory. It is

$$\frac{1}{r} = -\frac{M\kappa}{L^2}\left[1 + \sqrt{1 + \frac{2HL^2}{M\kappa^2}}\cos\left(\theta - \theta_0\right)\right], \tag{4.45}$$

where H is the Hamiltonian, which represents the conserved total energy, and L is the conserved angular momentum about the scattering center at O. The trajectory is symmetric about a line going outward from the scattering center at angle θ_0, because the cosine is an even function. The square root is called the *eccentricity* of the orbit ϵ. In the case where $\epsilon > 1$ the trajectory is a hyperbola, with asymptotes shown in Figure 4.4. In the case of an attractive force, this requires that the total energy H be sufficiently high that the particle is not bound.

Letting $r \to \infty$ after the scattering has taken place, we obtain

$$0 = 1 + \sqrt{1 + \frac{2HL^2}{M\kappa^2}} \cos(\vartheta - \theta_0). \tag{4.46}$$

It is evident from the figure that

$$2(\theta_0 - \vartheta) + \vartheta = \pi, \tag{4.47}$$

from which it follows that

$$\cos(\vartheta - \theta_0) = \cos\left(\frac{\vartheta}{2} - \frac{\pi}{2}\right) = \sin(\vartheta/2). \tag{4.48}$$

The conserved angular momentum $\mathbf{L}$ is given at any given point along the trajectory by

$$\mathbf{L} = \mathbf{r} \times \mathbf{p}. \tag{4.49}$$

Considering the incident particle far from the scattering center, we write

$$L = \lim_{r \to \infty} r\, p_0 \cdot \frac{b}{r} = b\sqrt{2MH}, \tag{4.50}$$

where $p_0 = \sqrt{2MH}$ is the initial momentum. Substituting this into (4.46), we find

$$\sin(\vartheta/2) = -\left(1 + \frac{4H^2 b^2}{\kappa^2}\right)^{-1/2}. \tag{4.51}$$

Solving for the impact parameter b, this leads to

$$b = \frac{\kappa}{2H} \cot(\vartheta/2). \tag{4.52}$$

Differentiating, we obtain

$$\frac{db}{d\vartheta} = -\frac{\kappa}{4H} \csc^2(\vartheta/2). \tag{4.53}$$

Substituting (4.52, 4.53) into (4.43), we obtain the result for the differential cross section as

$$\sigma(\vartheta) = \frac{\kappa^2}{16\,H^2\,\sin^4(\vartheta/2)}. \tag{4.54}$$

This is the dependence seen by Rutherford in the scattering of alpha particles by a gold foil, from which the nuclear model of the atom was originally deduced. It is therefore called the Rutherford scattering cross section. It is the same for both signs of κ, and is therefore independent of whether the scattering force is attractive or repulsive. It approaches infinity at zero scattering angle, and has a finite value for backscattering at $\vartheta = \pi$. Integrating over all solid angle, we form the *total cross section*. This is

$$\sigma_{tot} = 2\pi \int_0^{\pi} \sigma(\vartheta) \, \sin \vartheta \, d\vartheta. \qquad (4.55)$$

This is infinite for the case of Rutherford scattering. Physically, this means that the Coulomb potential effectively has infinite range.

We are now in a good position to consider quantum mechanical elastic scattering. This is the subject of the next three sections.

4.3   Integral expression of Schrödinger's equation

All relevant information about quantum mechanical scattering is contained in the differential cross section, which was defined in the preceding section. The cross section in turn depends on the wave function $\psi(\mathbf{x}, t)$, which is a solution of the time-dependent Schödinger equation (3.13) with appropriate boundary conditions. In this section we cast Schrödinger's equation in a form which will prove to be directly applicable to the scattering problem.

In the important special case where the electrostatic potential $\phi(\mathbf{x})$ has no explicit time dependence, the wave function $\psi(\mathbf{x}, t)$ takes on the separable form (3.18) where

$$\psi(\mathbf{x}, t) = u(\mathbf{x}) \, e^{-iHt/\hbar}, \qquad (4.56)$$

and H is the eigenvalue for the conserved total energy.

We now proceed to apply this formalism to the scattering problem. The spatial part $u(\mathbf{x})$ satisfies

$$\nabla^2 u(\mathbf{x}) + k^2\, u(\mathbf{x}) = \frac{2m}{\hbar^2}\, U(\mathbf{x})\, u(\mathbf{x}), \qquad (4.57)$$

where $U(\mathbf{x}) = q\phi(\mathbf{x})$ is the potential energy associated with the scattering center, and

$$k^2 = \frac{2mH}{\hbar^2}. \qquad (4.58)$$

This is recognizable as the Helmholtz equation. It is inhomogeneous, owing to the source term on the right-hand side. The ge-

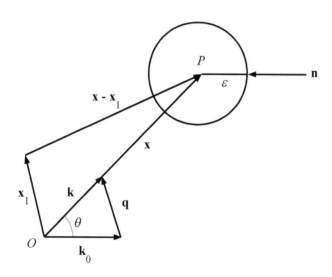

Figure 4.5: Geometry for scattering.

ometry is shown schematically in Figure 4.5. The scattering center is located at point O. The scattering is described by the potential energy $U(\mathbf{x}_1)$, assumed spherically symmetric about O. We seek the scattered wave function $u(\mathbf{x})$ at the observation point P, located a large distance $\mathbf{x}$ from O. The incident plane wave and the scattered spherical wave are described by the wave vectors $\mathbf{k}_0$ and $\mathbf{k}$, respectively. The angle θ between them is the scattering angle.

The difference between these outgoing and incoming wave vectors is $\mathbf{q} = \mathbf{k} - \mathbf{k}_0$. For elastic scattering the magnitudes k_0 and k are equal.

We postulate a point source at P, radiating an outgoing spherical wave, represented by the complex wave function

$$G(R) = \frac{1}{R} \exp(ikR). \tag{4.59}$$

This point source does not exist physically, but provides a mathematical aid to solve the problem. The amplitude $G(R)$ must satisfy the Helmholtz equation,

$$\nabla_R^2 G(R) + k^2 G(R) = 0 \tag{4.60}$$

everywhere except at P, where $G(R)$ has a singularity. To verify that this is the case, we express the Laplacian operator ∇^2 in spherical coordinates, leading to

$$\frac{1}{R} \frac{d^2}{dR^2} [RG(R)] + k^2 G(R) = 0. \tag{4.61}$$

The solution is immediately recognizable as

$$RG(R) = \exp(\pm ikR), \tag{4.62}$$

in agreement with (4.59) as required. We evaluate $G(R)$ at the field point $\mathbf{x}_1$ in Figure 4.5, where

$$R = |\mathbf{x} - \mathbf{x}_1|. \tag{4.63}$$

For notational purposes, we denote $G(R) = G(\mathbf{x}, \mathbf{x}_1)$ in the following, where

$$G(\mathbf{x}, \mathbf{x}_1) = \frac{1}{|\mathbf{x} - \mathbf{x}_1|} \exp(ik|\mathbf{x} - \mathbf{x}_1|) \tag{4.64}$$

represents the outgoing spherical wave. Multiplying (4.57) by $G(\mathbf{x}, \mathbf{x}_1)$, multiplying (4.60) by $u(\mathbf{x}_1)$, and subtracting the two equations, we obtain

$$G(\mathbf{x}, \mathbf{x}_1) \nabla_1^2 u(\mathbf{x}_1) - u(\mathbf{x}_1) \nabla_1^2 G(\mathbf{x}, \mathbf{x}_1) = \frac{2m}{\hbar^2} U(\mathbf{x}_1) G(\mathbf{x}, \mathbf{x}_1) u(\mathbf{x}_1). \tag{4.65}$$

We now proceed to integrate this over the entire space, excepting the small sphere about P, where G is singular. We make use of the divergence theorem to convert the volume integral on the left side to a surface integral over the entire surface S_1 surrounding the volume. This gives

$$\int_{S_1} dS_1 \left[G(\mathbf{x}, \mathbf{x}_1) \frac{\partial}{\partial n} u(\mathbf{x}_1) - u(\mathbf{x}_1) \frac{\partial}{\partial n} G(\mathbf{x}, \mathbf{x}_1) \right]$$

$$= \frac{2m}{\hbar^2} \int_{V_1} d^3\mathbf{x}_1\, U(\mathbf{x}_1)\, G(\mathbf{x}, \mathbf{x}_1)\, u(\mathbf{x}_1), \qquad (4.66)$$

where n denotes the outward normal to the surface S_1. The surface integral over S_1 consists of the sum of two contributions, namely, the small sphere S_ϵ of radius ϵ about P, and a large sphere S_∞ at infinty. Taking the small sphere first, we find

$$\int_{S_\epsilon} dS_1 \left[G(\mathbf{x}, \mathbf{x}_1) \frac{\partial}{\partial n} u(\mathbf{x}_1) - u(\mathbf{x}_1) \frac{\partial}{\partial n} G(\mathbf{x}, \mathbf{x}_1) \right] \to 4\pi\, u(\mathbf{x})$$

$$(4.67)$$

in the limit $\epsilon \to 0$, where the second term on the left predominates, and the first term becomes negligible. Considering the sphere S_∞ at infinity, we find

$$\int_{S_\infty} dS_1 \left[G(\mathbf{x}, \mathbf{x}_1) \frac{\partial}{\partial n} u(\mathbf{x}_1) - u(\mathbf{x}_1) \frac{\partial}{\partial n} G(\mathbf{x}, \mathbf{x}_1) \right]$$

$$\to \int_{S_\infty} \left(\frac{\partial u}{\partial n} - iku \right) R^2\, G(R)\, d\Omega, \qquad (4.68)$$

where $d\Omega$ is the element of solid angle. The right side vanishes, as long as

$$\left(\frac{\partial u}{\partial n} - iku \right) R \to 0 \qquad (4.69)$$

at infinity. This is, in fact, the case, where (4.69) is known as the Sommerfeld radiation condition [85].

We are thus left with an equation for $u(\mathbf{x})$ as follows:

$$u(\mathbf{x}) = \frac{m}{2\pi\hbar^2} \int d^3\mathbf{x}_1\, G(\mathbf{x}; \mathbf{x}_1)\, U(\mathbf{x}_1)\, u(\mathbf{x}_1), \qquad (4.70)$$

where $G(\mathbf{x}; \mathbf{x}')$ is the Green's function (4.64). This equation represents the main result of this section. It can be regarded as completely equivalent to the differential equation for $u(\mathbf{x})$ (4.57), which is the spatial part of Schrödinger's equation for the stationary-state case. We recall that $|u(\mathbf{x})|^2$ is the probability density that a single, precise measurement of the scattered particle position will find the particle at position $\mathbf{x}$. This represents the connection with experimental measurement. The equation (4.70) has the advantage of being more directly applicable to the scattering problem. We will make use of this in the following section.

4.4 Green's function solution for elastic scattering

We now turn our attention to the important special case of two-particle scattering in which the incident particle transfers negligible energy to the target particle. This process is known as *elastic* scattering. The central problem is to calculate the differential cross section $\sigma(\vartheta)$, which gives the scattered intensity at angle ϑ.

In the case where the incident and target particles each retain their same rest mass in the initial and final states, and we disregard internal degrees of freedom for each particle, the total energy is conserved for each particle individually in the CM frame. In the nonrelativistic limit, the two-body scattering is reduced to the equivalent one-body scattering. In this case a single particle with reduced mass M scatters from a fixed center, where M is given by (4.15)

$$M = \frac{m_1 \, m_2}{m_1 + m_2}, \tag{4.71}$$

where m_1 and m_2 are the rest masses of the incident and target particles, respectively.

In the case where $m_1 \ll m_2$, the CM moves slowly in the lab frame. Also, $M \approx m_1$. In this limit the incident particle transfers a small fraction of its energy to the target particle. The scattering can therefore be regarded as elastic in the lab frame, as well as in the CM frame. This is a fair approximation for a fast electron incident on an atomic nucleus, for example.

In the following analysis, we will calculate the differential cross-section $\sigma(\vartheta)$ for the equivalent one-body scattering in the CM frame. The geometry of the scattering is shown for the equivalent one-body problem in Figure 4.5. We assume a plane wave incident from the left, with wave vector $\mathbf{k}_0$ oriented along the positive z-axis. A scattering center is located at O, and an observation point at P, at position $\mathbf{x}$. A spherical wave with wave vector $\mathbf{k}$ emanates from the scattering center O. The polar scattering angle between the incident and scattered wave vectors $\mathbf{k}_0$ and $\mathbf{k}$ is ϑ.

We define the normalized incident wave function $u_0(\mathbf{x})$ as the plane wave

$$u_0(\mathbf{x}) = \frac{1}{\sqrt{V}} e^{i\mathbf{k}_0 \cdot \mathbf{x}}. \tag{4.72}$$

We assume the observation point P is located far from the scattering center. As such, the scattered wave $u(\mathbf{x})$ can be approximated by a spherical wave,

$$u(\mathbf{x}) = f(\mathbf{k}_0, \mathbf{k}) \frac{1}{\sqrt{V}} \frac{e^{ikr}}{r}, \tag{4.73}$$

where we define $r \equiv |\mathbf{x}|$, and the factor $f(\mathbf{k}_0, \mathbf{k})$ is called the *scattering amplitude*.

Strictly, the incident plane wave has infinite extent. However, in practice the incident beam is typically collimated, so that the incident and scattered waves do not interfere at the observation point P. We write the incident flux S_0 and the scattered flux S, respectively as

$$S_0 = |u_0(\mathbf{x})|^2 v_0, \qquad S = |u(\mathbf{x})|^2 v \tag{4.74}$$

where v_0 and v are the incident and scattered velocities, respectively. For the case of elastic scattering, $v = v_0$. It follows immediately that the differential cross section is (4.73, 4.74)

$$\sigma = |f(\mathbf{k}_0, \mathbf{k})|^2. \tag{4.75}$$

We define a total wave function $u_T(\mathbf{x}) = u_0(\mathbf{x}) + u(\mathbf{x})$ as the sum of the incident and scattered wave functions. This must satisfy Schrödinger's equation,

$$\left(\nabla^2 + k^2\right) u_T(\mathbf{x}) = \frac{2m}{\hbar^2} U(\mathbf{x})\, u_T(\mathbf{x}) \tag{4.76}$$

where $U(\mathbf{x})$ is the potential energy associated with the scattering center, and

$$k^2 = \frac{2mH}{\hbar^2} \tag{4.77}$$

where H is the continuous total energy eigenvalue associated with the state u_T. This is recognizable as the Helmholtz equation. It is inhomogeneous, owing to the source term on the right-hand side.

From the previous section, this equation can be expressed in integral form as

$$u(\mathbf{x}) = \frac{m}{2\pi\hbar^2} \int d^3\mathbf{x}_1\, G(\mathbf{x}, \mathbf{x}_1)\, U(\mathbf{x}_1)\, u_T(\mathbf{x}_1) \tag{4.78}$$

where $G(\mathbf{x}, \mathbf{x}_1)$ is the Green's function given by

$$G(\mathbf{x}, \mathbf{x}_1) = \frac{1}{|\mathbf{x} - \mathbf{x}_1|} \exp\left(ik|\mathbf{x} - \mathbf{x}_1|\right). \tag{4.79}$$

The scattering potential energy $U(\mathbf{x}_1)$ is appreciably different from zero over a very small region $\mathbf{x}_1$. As the observation point P is very far away, we assume $r \gg r_1$, where we define $r \equiv |\mathbf{x}|$ and $r_1 \equiv |\mathbf{x}_1|$. From the law of cosines,

$$\begin{aligned} |\mathbf{x} - \mathbf{x}_1|^2 &= r^2 + r_1^2 - 2\mathbf{x} \cdot \mathbf{x}_1 \\ &\approx r^2 \left(1 - 2\frac{\mathbf{x} \cdot \mathbf{x}_1}{r^2}\right). \end{aligned} \tag{4.80}$$

Taking the square root of both sides, and retaining only the two largest terms in the Taylor series expansion,

$$|\mathbf{x} - \mathbf{x}_1| \approx r - \frac{\mathbf{x} \cdot \mathbf{x}_1}{r}. \tag{4.81}$$

It follows that (4.78, 4.79, 4.81)

$$u(\mathbf{x}) \approx \frac{m}{2\pi\hbar^2} \frac{e^{ikr}}{r} \int d^3\mathbf{x}_1 \exp\left(-ik \frac{\mathbf{x} \cdot \mathbf{x}_1}{r}\right) U(\mathbf{x}_1) u_T(\mathbf{x}_1). \tag{4.82}$$

From the definition of the scattering amplitude f, it follows (4.73) that

$$f = \frac{m\sqrt{V}}{2\pi\hbar^2} \int d^3\mathbf{x}_1 e^{-i\mathbf{k}\cdot\mathbf{x}_1} U(\mathbf{x}_1) u_T(\mathbf{x}_1) \tag{4.83}$$

where we have defined the scattered wave vector as

$$\mathbf{k} \equiv k \frac{\mathbf{x}}{r}, \tag{4.84}$$

noticing that $\mathbf{x}/r$ is the unit vector in the direction of the scattering. This equation cannot be solved in closed form, because of the presence of the still unknown u_T under the integral. Therefore, we must seek a suitable approximation. To this end we replace u_T under the integral by the incident wave function u_0. This is known as the *first Born approximation*. It is justifiable when the scattering is relatively weak. In this approximation, we write (4.72, 4.83)

$$f(\mathbf{q}) = \frac{m}{2\pi\hbar^2} \int d^3\mathbf{x}_1 U(\mathbf{x}_1) e^{-i\mathbf{q}\cdot\mathbf{x}_1}, \tag{4.85}$$

where we have defined the difference vector $\mathbf{q}$ as

$$\mathbf{q} = \mathbf{k} - \mathbf{k}_0. \tag{4.86}$$

We recognize this as the Fourier transform of the scattering potential energy distribution $U(\mathbf{x}_1)$. As $\hbar\mathbf{k}_0$ and $\hbar\mathbf{k}$ represent the incident and scattered momenta, respectively, it follows that $\hbar\mathbf{q}$ is the momentum transferred in the collision. This expression for $f(\mathbf{q})$ is quite general, as we have not yet specified the precise form of the scattering potential energy $U(\mathbf{x}_1)$. It applies in many cases of elastic scattering, including high energy particle physics, and electron microscopy, to name just two.

We now turn to the important special case where $U(\mathbf{x})$ is spherically symmetric; i.e., $U(\mathbf{x}) = U(r)$. In spherical coordinates,

$$d^3\mathbf{x}_1 = r_1^2 \sin\theta_1 \, dr_1 \, d\theta_1 \, d\phi_1. \tag{4.87}$$

We assume for the present analysis that the scattering is azimuthally symmetric, in which case we immediately integrate over ϕ_1 to give (4.85, 4.87)

$$f(q) = \frac{m}{\hbar^2} \int_0^\infty dr_1 \, r_1^2 \, U(r_1) \int_0^\pi d\theta_1 \, \sin\theta_1 \, e^{-iqr_1 \cos\theta_1}. \tag{4.88}$$

Substituting $\cos\theta_1 \equiv \mu$, we find

$$\int_0^\pi d\theta_1 \, \sin\theta_1 \, e^{-iqr_1 \cos\theta_1} = \int_{-1}^1 d\mu \, e^{-iqr_1\mu}$$

$$= \frac{2}{qr_1} \sin(qr_1) \tag{4.89}$$

and (4.88, 4.89)

$$f(q) = \frac{2m}{\hbar^2 q} \int_0^\infty dr_1 \, r_1 \, U(r_1) \sin(qr_1). \tag{4.90}$$

This applies to any elastic scattering process for which the scattering potential energy is spherically symmetric.

We now study the special case where an incident particle of charge ze is elastically scattered by the screened Coulomb potential of a target atomic nucleus of charge Ze. This process represents the basic mechanism of contrast formation in a transmission electron microscope, for example. We now assume that the spherically symmetric potential $U(r_1)$ is represented by the screened Coulomb potential

$$U(r_1) = \frac{Zze^2}{4\pi\epsilon_0 \, r_1} \exp\left(-\alpha r_1\right). \tag{4.91}$$

Physically, this means that the bare charge of the scattering nucleus is screened by the electron charges of the target atom. This limits the spatial extent of the scattering region, compared with a bare, unscreened nuclear charge. Substituting, we find (4.90, 4.91)

$$f(q) = \frac{mZze^2}{2\pi\epsilon_0\hbar^2} \frac{1}{q} \int_0^\infty dr_1 \, e^{-\alpha r_1} \sin(qr_1) \qquad (4.92)$$

Making the substitutions

$$\xi \equiv qr_1, \qquad\qquad \beta \equiv \frac{\alpha}{q}, \qquad\qquad (4.93)$$

we obtain (4.92)

$$\begin{aligned}
f(q) &= \frac{mZze^2}{2\pi\epsilon_0\hbar^2} \frac{1}{q^2} \int_0^\infty d\xi \, e^{-\beta\xi} \sin\xi \\
&= \frac{mZze^2}{2\pi\epsilon_0\hbar^2} \frac{1}{q^2 + \alpha^2},
\end{aligned} \qquad (4.94)$$

where we have made use of

$$\int_0^\infty d\xi \, e^{-\beta\xi} \sin\xi = \frac{1}{1+\beta^2}. \qquad (4.95)$$

For an incident electron with $z = 1$, this takes the form

$$f(q) = \frac{4\pi Z}{137\,\lambda_C\,(q^2 + \alpha^2)}, \qquad (4.96)$$

where we have made use of

$$\frac{e^2}{4\pi\epsilon_0\hbar c} = \frac{1}{137}, \qquad (4.97)$$

and the Compton wavelength λ_C for the electron, defined by

$$\lambda_C = \frac{h}{mc} = 0.002435\,\text{nm}. \qquad (4.98)$$

The differential cross section σ is then given (4.75, 4.94) by

$$\sigma(q) = |f(q)|^2 = \left(\frac{mZze^2}{2\pi\epsilon_0\hbar^2}\right)^2 \frac{1}{(q^2 + \alpha^2)^2}. \qquad (4.99)$$

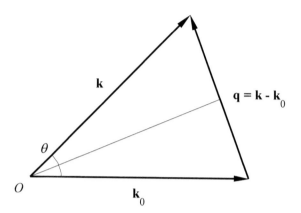

Figure 4.6: Momentum transfer and scattering angle.

The quantity q is related to scattering angle ϑ by

$$q = 2k \sin (\vartheta/2), \qquad (4.100)$$

as is evident from Figure 4.6. Substituting, we find (4.99, 4.100)

$$\sigma(\vartheta) = \left(\frac{mZze^2}{2\pi\epsilon_0\hbar^2}\right)^2 \frac{1}{\left[4k^2\sin^2(\vartheta/2) + \alpha^2\right]^2}. \qquad (4.101)$$

In the limit $\alpha \to 0$, we obtain (4.101)

$$\sigma(\vartheta) = \left[\frac{Zze^2}{16\pi\epsilon_0 H \sin^2(\vartheta/2)}\right]^2, \qquad (4.102)$$

where H is the total energy eigenvalue of the incident particle. This is equal to the kinetic energy of the particle outside of the scattering field. This is identical with the result for Rutherford scattering (4.44, 4.54) by an unscreened nucleus, derived previously from classical mechanics. This remarkable equivalence between the classical and quantum mechanical results only holds true for the scattering potential energy inversely proportional to the separation r.

We now proceed to integrate the differential cross section σ over all solid angles $d\Omega$, to obtain the *total cross section* σ_e for elastic scattering as follows:

$$\sigma_e = \int_{4\pi} \sigma(q)\, d\Omega, \tag{4.103}$$

where $d\Omega = 2\pi \sin\vartheta\, d\vartheta$. We notice that (4.100)

$$\frac{d\Omega}{dq} = \frac{d\Omega}{d\vartheta}\frac{d\vartheta}{dq} = 2\pi \sin\vartheta \frac{1}{k\cos(\vartheta/2)} = \frac{2\pi q}{k^2}, \tag{4.104}$$

where $0 \le q \le 2k$. It follows that (4.99, 4.103, 4.104)

$$\begin{aligned}\sigma_e &= \frac{2\pi}{k^2}\int_0^{2k}\sigma(q)\,q\,dq \\ &= \left(\frac{mZze^2}{2\pi\epsilon_0\hbar^2}\right)^2 \frac{4\pi}{\alpha^2\,(4k^2+\alpha^2)}. \end{aligned} \tag{4.105}$$

In the limit $\alpha \to 0$, where the nuclear charge is unscreened, the total cross section becomes infinite. The Coulomb force is therefore said to have infinite range. Physically, α represents the reciprocal of the radius of the atomic electron cloud. It is approximately

$$\alpha \approx \frac{Z^{1/3}}{a_0} \tag{4.106}$$

where a_0 is the Bohr radius of the hydrogen atom given by

$$a_0 = \frac{4\pi\epsilon_0\hbar^2}{e^2 m} = \frac{137\,\lambda_C}{2\pi} = 0.0531 \text{ nm}. \tag{4.107}$$

For incident energy $H > 1$ KeV, we have $4k^2 \gg \alpha^2$, in which case (4.77, 4.105)

$$\begin{aligned}\sigma_e &\approx \left(\frac{mZze^2}{2\pi\epsilon_0\hbar^2}\right)^2 \frac{\pi}{k^2\alpha^2} \\ &= \frac{2\pi m}{H}\left(\frac{mZze^2}{2\pi\epsilon_0\hbar^2}\right)^2. \end{aligned} \tag{4.108}$$

Taking

$$H = \tfrac{1}{2}m\beta^2 c^2, \qquad\qquad \beta = v/c, \qquad\qquad z = 1, \tag{4.109}$$

we obtain a useful approximation for the total elastic cross section
for an incident electron (4.108, 4.109) as

$$\sigma_e = \frac{Z^{4/3}}{\beta^2} \frac{\lambda_C^2}{\pi} = 1.9 \times 10^{-6} \frac{Z^{4/3}}{\beta^2} \text{ nm}^2. \tag{4.110}$$

Knowing the total cross section, we can now estimate the mean
free path μ_e which an electron travels between scattering events.
This is

$$\mu_e = \frac{1}{N\sigma_e} = \frac{A}{N_0 \rho \sigma_e}, \tag{4.111}$$

where N = number of scattering centers per unit volume, A =
atomic number, N_0 = Avagadro's number, and ρ = mass density. For 100 KeV electrons incident on silicon ($Z = 14$, $A = 28$
gm/mole, $\rho = 2.4$ gm/cm^3), we find $\mu_e = 93$ nm for the mean free
path.

For fast electrons incident on a film of thickness of the order of
the mean free path, the average scattering angle is quite small. In
this case, we can approximate

$$\sin\frac{\vartheta}{2} \approx \frac{\vartheta}{2}. \tag{4.112}$$

The differential cross section $\sigma(\vartheta)$ is then approximately given
(4.101, 4.112) by

$$\sigma(\vartheta) = \left(\frac{Zze^2}{4\pi\epsilon_0 H}\right)^2 \frac{1}{(\vartheta^2 + \vartheta_W^2)^2} \tag{4.113}$$

where we have defined (4.106)

$$\vartheta_W = \frac{\alpha}{k} = \frac{Z^{1/3}\lambda}{2\pi a_0} \tag{4.114}$$

where $\lambda = h/p$ is the particle wavelength. The angle ϑ_W is called
the *Wentzel screening angle*. For 100 KeV electrons incident on
silicon ($Z = 14$, $\lambda = 0.0037$ nm), we find $\vartheta_W = 0.027$ rad, consistent with our assumption of small angles. The total cross section

is (4.101, 4.103, 4.112)

$$
\sigma_e = \int_{4\pi} \sigma \, d\Omega = \left(\frac{Zze^2}{4\pi\epsilon_0 H} \right)^2 \pi \int_0^\infty \frac{2\vartheta \, d\vartheta}{(\vartheta^2 + \vartheta_W^2)^2}
$$

$$
= \left(\frac{Zze^2}{4\pi\epsilon_0 H} \right)^2 \frac{\pi}{\vartheta_W^2}. \tag{4.115}
$$

For 100 KeV electrons incident on silicon, this yields $\sigma_e = 1.8 \times 10^{-4}$ nm^2. We can form an angular distribution which is normalized to unity as $\sigma(\vartheta)/\sigma_e$. This is

$$
\sigma_1(\vartheta) = \frac{\vartheta_W^2}{\pi} \frac{1}{(\vartheta^2 + \vartheta_W^2)^2} \tag{4.116}
$$

where

$$
2\pi \int_0^\infty \sigma_1(\vartheta) \, \vartheta \, d\vartheta = 1. \tag{4.117}
$$

The normalized distribution σ_1 will prove useful in the theory of small angle plural scattering.

4.5 Perturbation theory

At this point we describe what happens when a quantum mechanical system experiences a small perturbation from its initial undisturbed state. This will provide a very useful mathematical tool to further understand scattering.

We consider a general system, which is described by a Hamiltonian operator $\hat{H}_0$ satisfying

$$
\hat{H}_0 \, \psi(\mathbf{x}, t) = i\hbar \frac{\partial}{\partial t} \psi(\mathbf{x}, t), \tag{4.118}
$$

where $\psi(\mathbf{x}, t)$ is the eigenfunction, and the Hamiltonian $\hat{H}_0$ is assumed to have no explicit time dependence. The eigenfunction for the jth state is given by

$$
\psi_j(\mathbf{x}, t) = u_j(\mathbf{x}) \, e^{-iH_j t/\hbar}. \tag{4.119}
$$

A linear superposition of eigenfunctions $\psi_j(\mathbf{x}, t)$ yields the state function

$$\Psi_0(\mathbf{x}, t) = \sum_j a_j \, u_j(\mathbf{x}) \, e^{-iH_j t/\hbar}, \qquad (4.120)$$

where $a_j = const$, and $|a_j|^2$ is the probability that a single, precise measurement of the total energy will yield the eigenvalue H_j.

This description is quite general, and applies to a variety of quantum mechanical systems. As examples the system might consist of

- a free particle,

- a particle in a general electromagnetic potential,

- a particle in the presence of the screened Coulomb potential of a target nucleus,

- a particle in the presence of an atom consisting of a nucleus and a cloud of electrons.

We now introduce a perturbation, by assuming a Hamiltonian $\hat{H}$ consisting of two terms,

$$\hat{H} = \hat{H}_0 + \hat{H}_1(t). \qquad (4.121)$$

The first term $\hat{H}_0$ is the unperturbed Hamiltonian in the absence of any interaction between the constituent parts of the system. The second term $\hat{H}_1$ is a perturbation representing the interaction. In general this perturbation depends on the time t.

Since the unperturbed eigenfunctions $\psi_j(\mathbf{x}, t)$ form an orthonormal set, it is always possible to expand the perturbed state function $\Psi(\mathbf{x}, t)$ as a linear combination of the unperturbed eigenfunctions. Thus

$$\Psi(\mathbf{x}, t) = \sum_j a_j(t) \, u_j(\mathbf{x}) \, e^{-iH_j t/\hbar}, \qquad (4.122)$$

where the coefficients $a_j(t)$ are now considered to depend on the time t, owing to the time dependence of the perturbation $\hat{H}_1(t)$.

Substituting this into

$$(\hat{H}_0 + \hat{H}_1)\,\Psi(\mathbf{x},t) = i\hbar \frac{\partial}{\partial t}\,\Psi(\mathbf{x},t) \tag{4.123}$$

and subtracting out the unperturbed terms, we find

$$\sum_j a_j\,(\hat{H}_1 u_j)\,e^{-iH_j t/\hbar} = i\hbar \sum_j \frac{da_j}{dt}\,u_j(\mathbf{x})\,e^{-iH_j t/\hbar}. \tag{4.124}$$

Multiplying from the left by $\bar{u}_i(\mathbf{x})$ and integrating over the volume,

$$i\hbar \frac{d}{dt}\,a_i(t) = \sum_j a_j(t)\,e^{i(H_i - H_j)t/\hbar} \int d^3x\,\bar{u}_i(\mathbf{x})\,[\,\hat{H}_1\,u_j(\mathbf{x})\,], \tag{4.125}$$

where we have made use of the orthonormality of the u_j, namely

$$\int d^3x\,\bar{u}_i(\mathbf{x})\,u_j(\mathbf{x}) = \delta_{ij}. \tag{4.126}$$

For brevity we make use of the Dirac notation as follows:

$$\int d^3x\,\bar{u}_i(\mathbf{x})\,[\,\hat{H}_1\,u_j(\mathbf{x})\,] = \left\langle i|\hat{H}_1|j\right\rangle. \tag{4.127}$$

We further abbreviate

$$H_i - H_j = H_{ij}. \tag{4.128}$$

In this notation we have

$$i\hbar \frac{d}{dt}\,a_i(t) = \sum_j a_j(t)\,e^{iH_{ij}t/\hbar}\left\langle i|\hat{H}_1|j\right\rangle. \tag{4.129}$$

This equation describes the time evolution of the amplitude $a_i(t)$ in the presence of the perturbation. It is exact, since no approximation has been made to this point.

We now introduce several approximating assumptions as follows:

- Initially, only one state is populated, and all other states are unpopulated. Mathematically, $a_j(0) = 1$ for one specific value of the index j, and $a_k(0) = 0$ for all other values $k \neq j$.

- The interaction begins instantaneously at time $t = 0$, and remains constant thereafter. In this respect we regard $\hat{H}_1$ as time independent after $t = 0$.

- The change in each $a_j(t)$ over time is small throughout the time of interaction. This is equivalent to the perturbation being weak over the time scale of interest. Mathematically, $a_j(t) \approx a_j(0)$ for all j and all t.

Based on these assumptions,

$$\frac{d}{dt} a_i(t) \approx \frac{1}{i\hbar} \left\langle i|\hat{H}_1|j \right\rangle e^{iH_{ij}t/\hbar}. \qquad (4.130)$$

The index i runs over a multiplicity of final states with energy H_i close to H_j. Integrating over time,

$$
\begin{aligned}
a_i(t) &= \frac{1}{i\hbar} \left\langle i|\hat{H}_1|j \right\rangle \int_0^t e^{iH_{ij}t/\hbar} dt \\
&= -\frac{1}{H_{ij}} \left\langle i|\hat{H}_1|j \right\rangle \left(e^{iH_{ij}t/\hbar} - 1 \right) \\
&= -\frac{1}{H_{ij}} \left\langle i|\hat{H}_1|j \right\rangle \exp\left(\frac{iH_{ij}t}{2\hbar} \right) \cdot 2i \sin\left(\frac{H_{ij}t}{2\hbar} \right).
\end{aligned}
$$

$$(4.131)$$

The probability $|a_i(t)|^2$ of finding the final state i at time t is then

$$|a_i(t)|^2 = \left| \frac{2}{H_{ij}} \left\langle i|\hat{H}_1|j \right\rangle \sin\left(\frac{H_{ij}t}{2\hbar} \right) \right|^2. \qquad (4.132)$$

The probability of transition from the initial state j to *all* final states is found by summing over the final states i,

$$P(t) = \sum_i |a_i(t)|^2. \qquad (4.133)$$

In the important case of an unbound system where the final states i approach a continuum, this becomes

$$P(t) = \int_{-\infty}^{\infty} dH_i \, \rho(H_i) \, |a_i(t)|^2, \qquad (4.134)$$

where $\rho(H_i)$ is the density of states with respect to energy. Assuming all final energies H_i are close to the initial energy H_j (weak perturbation), we can approximate

$$H_i \approx H_j \approx H. \tag{4.135}$$

Additionally, we approximate the matrix element by a single constant value

$$\left\langle i|\hat{H}_1|j \right\rangle \approx \langle H \rangle. \tag{4.136}$$

As a result, we can bring $\langle H \rangle$ and $\rho(H)$ outside the integral. Substituting,

$$P(t) = 4 \langle H \rangle^2 \rho(H) \int_{-\infty}^{\infty} \frac{dH_{ij}}{H_{ij}^2} \sin^2 \left(\frac{H_{ij}t}{2\hbar} \right). \tag{4.137}$$

Making the substitution

$$\xi \equiv \frac{H_{ij}t}{2\hbar}, \tag{4.138}$$

we find

$$P(t) = \frac{2\pi}{\hbar} \rho(H) \langle H \rangle^2 t, \tag{4.139}$$

where we have made use of the integral

$$\int_{-\infty}^{\infty} \frac{d\xi}{\xi^2} \sin^2 \xi = \pi. \tag{4.140}$$

The transition rate from a single initial state to all final states is then

$$\frac{dP}{dt} = \frac{2\pi}{\hbar} \rho(H) \langle H \rangle^2, \tag{4.141}$$

where the transition rate is the probability per unit time for the transition from a single initial state to *all* available final states. This result is quite general, in that it applies to many diverse phenomena. It is called the *golden rule* of perturbation theory. In the following sections we will proceed to apply it to the scattering problem.

4.6 Perturbation solution for elastic scattering

We now proceed to apply the equation (4.141) to the problem of elastic scattering. We assume that the initial state corresponds to an incident plane wave, where the free-particle eigenfunction $u_0(\mathbf{x})$ is given by

$$u_0(\mathbf{x}) = \frac{1}{\sqrt{V}} \, e^{i\mathbf{k}_0 \cdot \mathbf{x}}, \tag{4.142}$$

where V is the volume, $\mathbf{k}_0$ is the incident wave vector, and $\hbar\mathbf{k}_0$ is the incident momentum. The final state at a large distance from the scattering center is a plane wave given by

$$u(\mathbf{x}) = \frac{1}{\sqrt{V}} \, e^{i\mathbf{k} \cdot \mathbf{x}}, \tag{4.143}$$

where $\mathbf{k}$ is the scattered wave vector, and $\hbar\mathbf{k}$ is the momentum after scattering of the incident particle. In the initial and final states the particle is assumed to be far outside the region of scattering, hence the free-particle eigenfunctions.

The unperturbed Hamiltonian $\hat{H}_0$ is then the free particle Hamiltonian, and the perturbation Hamiltonian $\hat{H}_1$ is the scattering potential energy

$$\hat{H}_1 = U(\mathbf{x}). \tag{4.144}$$

We take the origin of coordinates $\mathbf{x}$ to coincide with the scattering center of the equivalent one-body problem. In the case of an electron incident on an atom, the origin coincides with the position of the atomic nucleus.

The matrix element $\langle i|\hat{H}_1|j\rangle$ is then given by

$$\langle i|\hat{H}_1|j\rangle = \frac{1}{V} \int d^3\mathbf{x} \, U(\mathbf{x}) \, e^{-i\mathbf{q}\cdot\mathbf{x}}, \tag{4.145}$$

where we have defined the difference vector $\mathbf{q} \equiv \mathbf{k} - \mathbf{k}_0$. The quantity $\hbar\mathbf{q}$ is the momentum transferred in the collision.

The density of states with respect to energy was found earlier (3.73) to be

$$\rho(H) = \frac{4\pi m V}{h^3} \sqrt{2m^3 H}, \tag{4.146}$$

where, for a free particle,

$$H = \frac{\hbar^2 k^2}{2m}. \tag{4.147}$$

Substituting, this is equivalent to

$$\rho = \frac{2mkV}{h^2}. \tag{4.148}$$

The transition rate from the initial state to *all* final states is (4.141, 4.145, 4.148)

$$\frac{dP}{dt} = \frac{mk}{\pi \hbar^3 V} \left| \int d^3x \, U(\mathbf{x}) \, e^{-i\mathbf{q}\cdot\mathbf{x}} \right|^2. \tag{4.149}$$

This represents the probability per unit time of scattering into *all* solid angles. For the differential cross section we seek the number of particles per unit time scattered into a *particular* solid angle element $d\Omega$. This is

$$\frac{dN}{dt} = \frac{dP}{dt} \frac{d\Omega}{4\pi}. \tag{4.150}$$

The definition of the differential cross section σ was given earlier by (4.39)

$$\frac{dN}{dt} = \sigma \, S_0 \, d\Omega, \tag{4.151}$$

where S_0 is the incident flux (particles per unit time per unit transverse area) given by

$$S_0 = |u_0(\mathbf{x})|^2 \, v_0 = \frac{\hbar k_0}{Vm}. \tag{4.152}$$

For elastic scattering, $k_0 = k$. That is, the *magnitude* of the momentum and therefore the wave vector is unchanged by the scattering. Solving for the differential cross section σ, we obtain

$$\sigma(\mathbf{q}) = \left| \frac{m}{2\pi\hbar^2} \int d^3x \, U(\mathbf{x}) \, e^{-i\mathbf{q}\cdot\mathbf{x}} \right|^2 \equiv |f(\mathbf{q})|^2, \tag{4.153}$$

where $f(\mathbf{q})$ is the scattering amplitude derived previously (4.85). This is the main result of this section.

As a reminder, this applies to an arbitrary scattering potential $U(\mathbf{x})$. Mathematically, the equation (4.153) teaches that the scattering amplitude is the Fourier transform of the scattering potential. In practical terms, the angular distribution of the scattering is directly measurable, and represents a sensitive probe into the detailed form of the scattering potential.

We have succeeded in reproducing the earlier result by the independent use of perturbation theory as an alternative approach. From this we deduce that the approximations made here coincide with the first Born approximation.

4.7 Inelastic scattering of a particle by a target atom

A scattering event which transfers energy from the incident particle to the target material is called *inelastic* scattering. The transferred energy can be manifest in a variety of secondary processes. These include emission of a photon, Auger electron, or ionization electron from a target atom. Alternatively they include collective excitation of the conduction band electron gas known as a *plasmon*, or of the target lattice as a *phonon*. Analysis of the energy lost by the primary particle provides important information about the chemical and physical composition of the target. At very high incident energy, various elementary particles can be created. Inelastic scattering is quite complicated, and the subject of an enormous literature.

In this study we confine our attention to the primary energy transfer, without considering the multiplicity of secondary processes. In one example, a massive incident particle transfers energy to cause excitation or ionization of the electrons of a target atom. In a second example, the passage of a fast charged particle causes instantaneous polarization of a dielectric medium. The central problem is to calculate the scattering cross section in each case.

We assume an incident particle of mass m and charge ze, where z is an integer. The target is a single atom with atomic number Z. In a single collision the incident particle transfers a small fraction of its energy to the target atom, initially in its ground state. As a result, the atom is excited to a higher energy state. In principle this can include ionization of the atom. The scattered particle then exits to a final free-particle state with reduced energy and momentum. The following analysis closely follows the classic paper of Bethe [5], which is based on the same perturbation-theoretical approach described above. The reader is referred to Egerton [24], who places the material in the context of the considerable body of subsequent work by others.

According to the foregoing analysis, calculation of the differential scattering cross section σ is reduced to finding the appropriate matrix element $\langle i|\hat{H}_1|j\rangle$ between the initial and final states of the system. In this case, the system consists of a scattering particle and a single target atom. The origin of coordinates coincides with the nucleus of the target atom. The instantaneous position of the scattering particle we denote by $\mathbf{x}$, and the instantaneous positions of the Z atomic electrons we denote by $(\mathbf{x}_1, \ldots, \mathbf{x}_Z)$.

The initial and final eigenfunctions, respectively, can be represented by

$$u_0(\mathbf{x}; \mathbf{x}_1, \ldots \mathbf{x}_Z) = \frac{1}{\sqrt{V}} e^{i\mathbf{k}_0 \cdot \mathbf{x}} \mathcal{U}_0(\mathbf{x}_1, \ldots, \mathbf{x}_Z)$$

$$u_n(\mathbf{x}; \mathbf{x}_1, \ldots \mathbf{x}_Z) = \frac{1}{\sqrt{V}} e^{i\mathbf{k} \cdot \mathbf{x}} \mathcal{U}_n(\mathbf{x}_1, \ldots, \mathbf{x}_Z), \quad (4.154)$$

where the subscripts 0 and n refer to the ground state and the nth excited states of the atom, respectively. The quantities $\mathcal{U}_0$ and $\mathcal{U}_n$

represent the spatial wave function of the atom before and after the collision. The time dependence has been integrated out in the perturbation-theoretical approach described above.

The scattering particle approaches from a large distance with incident wave vector $\mathbf{k}_0$, and exits with scattered wave vector $\mathbf{k}$. In this inelastic scattering case, the incident and scattered wave vectors differ in both magnitude and direction. The eigenfunctions u_0 and u_n represent solutions to the Schrödinger equation, where one must be careful to include the dependence on all $3(Z+1)$ spatial degrees of freedom, here labeled $(\mathbf{x}; \mathbf{x}_1, \ldots, \mathbf{x}_Z)$.

The matrix element is given by

$$\left\langle i|\hat{H}_1|j\right\rangle = \int \ldots \int \bar{u}_n \, U \, u_0 \, d^3x \prod_{j=1}^{Z} d^3x_j, \qquad (4.155)$$

where $U(\mathbf{x}; \mathbf{x}_1, \ldots \mathbf{x}_Z)$ is the potential energy arising from the Coulomb interaction. This is

$$U(\mathbf{x}; \mathbf{x}_1, \ldots, \mathbf{x}_Z) = \frac{e^2 z}{4\pi\epsilon_0} \left(\frac{Z}{|\mathbf{x}|} - \sum_{j=1}^{Z} \frac{1}{|\mathbf{x} - \mathbf{x}_j|} \right). \qquad (4.156)$$

The first term in large parentheses represents the interaction between the scattering particle and the bare atomic nucleus, and the second term represents the sum of interactions between the scattering particle and the atomic electrons.

The matrix element takes the form

$$\left\langle i|\hat{H}_1|j\right\rangle = \frac{1}{V} \int \ldots \int U \, e^{-i\mathbf{q}\cdot\mathbf{x}} \bar{\mathcal{U}}_n \, \mathcal{U}_0 \, d^3x \prod_{j=1}^{Z} d^3x_j, \qquad (4.157)$$

where $\mathbf{q} = \mathbf{k} - \mathbf{k}_0$. Following Bethe [5] we perform the integral over d^3x first. This integral is of the form

$$\int \frac{1}{|\mathbf{x} - \mathbf{x}_j|} e^{-i\mathbf{q}\cdot\mathbf{x}} \, d^3x = -\frac{4\pi}{q^2} e^{-i\mathbf{q}\cdot\mathbf{x}_j}. \qquad (4.158)$$

The truth of this equation can easily be established by applying the Laplacian operator $\nabla^2_{x_j}$ to both sides, where the subscript denotes differentiation with respect to the coordinates $\mathbf{x}_j$. Taking the Laplacian inside the integral on the left side, we make use of

$$\nabla^2_{x_j} \frac{1}{|\mathbf{x} - \mathbf{x}_j|} = -4\pi\,\delta(\mathbf{x} - \mathbf{x}_j), \qquad (4.159)$$

which is well-known from electrostatic potential theory [48]. Using the property of the delta-function, both sides are equal to $e^{-i\mathbf{q}\cdot\mathbf{x}_j}$, thus establishing the identity. As a special case we have

$$\int \frac{1}{|\mathbf{x}|}\,e^{-i\mathbf{q}\cdot\mathbf{x}}\,d^3x = -\frac{4\pi}{q^2}. \qquad (4.160)$$

The matrix element is reduced to

$$\langle i|\hat{H}_1|j\rangle = \frac{e^2 z}{\epsilon_0 q^2 V} \int \cdots \int \left(-Z + \sum_{j=1}^{Z} e^{-i\mathbf{q}\cdot\mathbf{x_j}} \right) \cdot \bar{\mathcal{U}}_n \mathcal{U}_0 \cdot \prod_{j=1}^{Z} d^3 x_j,$$
$$(4.161)$$

where the integral is now only over the coordinates of the Z atomic electrons $\mathbf{x}_j$. Making use of the orthonormality of the set $\mathcal{U}_n(\mathbf{x}_1, \ldots, \mathbf{x}_Z)$ this further reduces to

$$\begin{aligned}
\langle i|\hat{H}_1|j\rangle &= -\frac{e^2 z Z}{\epsilon_0 q^2 V}\,\delta_{n0} \\
&+ \frac{e^2 z}{\epsilon_0 q^2 V} \int \cdots \int \left(\sum_{j=1}^{Z} e^{-i\mathbf{q}\cdot\mathbf{x_j}} \right) \cdot \bar{\mathcal{U}}_n \mathcal{U}_0 \cdot \prod_{j=1}^{Z} d^3 x_j.
\end{aligned}$$
$$(4.162)$$

At this point we define a dimensionless quantity $\varepsilon_n(\mathbf{q})$ given by

$$\varepsilon_n(\mathbf{q}) = -Z\,\delta_{n0} + \int \cdots \int \left(\sum_{j=1}^{Z} e^{-i\mathbf{q}\cdot\mathbf{x_j}} \right) \cdot \bar{\mathcal{U}}_n \mathcal{U}_0 \cdot \prod_{j=1}^{Z} d^3 x_j, \quad (4.163)$$

where ε_n is a property of the target atom in the nth excited state. The first term represents the elastic scattering and the remainder represents the inelastic scattering. The matrix element is then

$$\langle H\rangle = \langle i|\hat{H}_1|j\rangle = \frac{e^2 z}{\epsilon_0\, q^2 V}\,\varepsilon_n. \qquad (4.164)$$

The transition rate is given by

$$\frac{dP}{dt} = \frac{2\pi}{\hbar} \rho \langle H \rangle^2,$$ (4.165)

where the density of states of the scattered particle is

$$\rho = \frac{2mVk}{h^2}.$$ (4.166)

The number of particles per unit time scattered into a solid angle element $d\Omega$ is given by

$$\frac{dN}{dt} = \frac{dP}{dt} \cdot \frac{d\Omega}{4\pi} = \sigma(q) \, S_0 \, d\Omega,$$ (4.167)

where S_0 is the incident intensity given by

$$S_0 = \frac{\hbar k_0}{mV}.$$ (4.168)

The central problem of this section is to calculate the differential cross section $\sigma(q)$. This is

$$\sigma(q) = \frac{\rho \langle H \rangle}{2 \hbar S_0}.$$ (4.169)

Substituting, we obtain the result

$$\sigma_n(q) = \left(\frac{m \, e^2 z}{2\pi\epsilon_0 \, \hbar^2 q^2} \right)^2 \frac{k}{k_0} \, | \, \varepsilon_n(\mathbf{q}) \, |^2,$$ (4.170)

where the subscript n indicates that the target atom is excited to the nth state. Energy conservation dictates that

$$E_0 + \frac{\hbar^2 k_0^2}{2m} = E_n + \frac{\hbar^2 k^2}{2m},$$ (4.171)

where E_0 and E_n are the ground state and nth excited state energy levels, respectively. The final state can consist of atomic excitation or ionization.

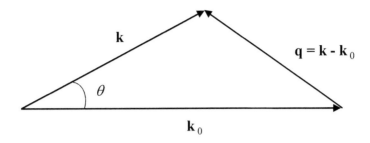

Figure 4.7: Wave vectors for inelastic scattering.

We recall that $\mathbf{k_0}$, $\mathbf{k}$, and $\mathbf{q}$ are related by

$$\mathbf{q} = \mathbf{k} - \mathbf{k_0}. \tag{4.172}$$

This is shown in Figure 4.7, where θ is the scattering angle. The various magnitudes are related by

$$q^2 = k^2 + k_0^2 - 2\,k\,k_0\,\cos\theta. \tag{4.173}$$

Taking the differential of both sides, we have

$$q\,dq = k\,k_0\,\sin\theta\,d\theta. \tag{4.174}$$

The solid angle element $d\Omega$ is given by

$$d\Omega = 2\pi\,\sin\theta\,d\theta. \tag{4.175}$$

Substituting, this leads to a differential form for the inelastic scattering cross section as

$$\sigma_n(q)\,d\Omega = \left(\frac{m\,e^2 z}{\epsilon_0\,\hbar^2}\right)^2 \frac{1}{2\pi\,k_0^2}\,|\,\varepsilon_n(\mathbf{q})\,|^2\,\frac{dq}{q^3}. \tag{4.176}$$

Integrating both sides over all possible values, we obtain the total cross section. This form shows that the total cross section for inelastic scattering is inversely proportional to the energy of the incident particle. Elastic scattering has the same inverse dependence on incident energy. For electron scattering the ratio of the total cross section σ_i for inelastic scattering divided by the total cross section σ_e for elastic scattering is given [24] by

$$\frac{\sigma_i}{\sigma_e} \approx \frac{20}{Z}. \tag{4.177}$$

Following Bethe [5], the momentum transfer q is related to the scattering angle θ by

$$q^2 \approx \theta^2 + \theta_E^2, \tag{4.178}$$

where θ_E is defined as

$$\theta_E = \frac{m\,\Delta\bar{E}}{\hbar^2 k_0^2}, \tag{4.179}$$

and $\Delta\bar{E}$ is the average energy loss per collision, and is in the range of a few eV to a few tens of eV, depending on the target material. Given that the scattering is from the electron cloud of the target atom, the predominant scattering angles for a fast incident particle are small, in the range of 1 mrad.

4.8   Slowing of a charged particle in a dielectric medium

When a fast charged particle passes through a solid, it interacts electromagnetically with many atoms simultaneously. In addition conduction band electrons are delocalized, with nonzero probability density over a region many atomic diameters in size. The incident particle interacts collectively with the electrons of the target material. It is worthwhile to consider the interaction in a classical approximation. This section closely follows the analysis by Landau and Lifshitz [56]. This material is described in more detail, and placed in the context of the earlier work of others by Egerton [24].

We consider a charged particle with velocity **v** passing through an infinite medium with complex dielectric coefficient $\epsilon(\omega)$. Here ω is the temporal angular frequency of the electromagnetic field of the particle. This presumes that the electromagnetic field is amenable to Fourier analysis. This is shown schematically in Figure 4.8. The

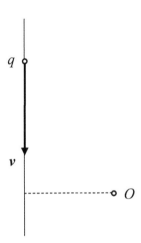

Figure 4.8: Particle passing through a dielectric medium.

particle has charge q and velocity **v**. An electromagnetic field is experienced at the observation point O due to the passing particle.

We adopt the nonrelativistic approximation, in which magnetic effects are negligible. The instantaneous electrostatic potential $\varphi(\mathbf{x})$ evaluated at any position $\mathbf{x}$ obeys Poisson's equation,

$$\nabla^2 \varphi(\mathbf{x}) = -\frac{\rho(\mathbf{x})}{\epsilon}. \tag{4.180}$$

The charge density ρ is due to the particle, and is given by

$$\rho(\mathbf{x}) = q\,\delta(\mathbf{x} - \mathbf{v}t). \tag{4.181}$$

The potential $\varphi(\mathbf{x})$ can be expressed as a Fourier integral,

$$\varphi(\mathbf{x}) = \int d^3k\, \tilde{\varphi}(\mathbf{k})\, e^{i\mathbf{k}\cdot\mathbf{x}}. \tag{4.182}$$

Applying the Laplacian operator to both sides, we obtain

$$\nabla^2 \varphi(\mathbf{x}) = -\int d^3k \, k^2 \, \tilde{\varphi}(\mathbf{k}) \, e^{i\mathbf{k}\cdot\mathbf{x}}. \qquad (4.183)$$

Separately, the delta function has the integral representation

$$\delta(\mathbf{x} - \mathbf{v}t) = \frac{1}{(2\pi)^3} \int d^3k \, e^{i\mathbf{k}\cdot(\mathbf{x}-\mathbf{v}t)}. \qquad (4.184)$$

Substituting into Poisson's equation above, we obtain

$$\tilde{\varphi}(\mathbf{k}) = \frac{q}{(2\pi)^3 \, k^2 \, \epsilon(\mathbf{k} \cdot \mathbf{v})} \, e^{i(\mathbf{k}\cdot\mathbf{v})t}. \qquad (4.185)$$

The electric field $\mathbf{E}(\mathbf{x})$ is given by

$$\begin{aligned} \mathbf{E}(\mathbf{x}) &= -\nabla \varphi(\mathbf{x}) \\ &= -\int d^3k \, \tilde{\varphi}(\mathbf{k}) \left(i\mathbf{k} \, e^{i\mathbf{k}\cdot\mathbf{x}} \right). \end{aligned} \qquad (4.186)$$

Separately, the electric field $\mathbf{E}(\mathbf{x})$ can be expressed as a Fourier integral,

$$\mathbf{E}(\mathbf{x}) = \int d^3k \, \tilde{\mathbf{E}}(\mathbf{k}) \, e^{i\mathbf{k}\cdot\mathbf{x}}. \qquad (4.187)$$

Substituting, we obtain

$$\begin{aligned} \tilde{\mathbf{E}}(\mathbf{k}) &= -i\mathbf{k} \, \tilde{\varphi}(\mathbf{k}) \\ &= -\frac{i\mathbf{k}q}{(2\pi)^3 \, k^2 \, \epsilon(\mathbf{k} \cdot \mathbf{v})} \, e^{i(\mathbf{k}\cdot\mathbf{v})t}. \end{aligned} \qquad (4.188)$$

Performing the inverse Fourier transform, and evaluating the electric field at the particle position $\mathbf{x} = \mathbf{v}t$, we obtain

$$\mathbf{E}(\mathbf{v}t) = -\frac{iq}{(2\pi)^3} \int d^3k \, \frac{\mathbf{k}}{k^2 \, \epsilon(\mathbf{k} \cdot \mathbf{v})}, \qquad (4.189)$$

where the exponential factors cancel. The force $\mathbf{F}$ on the particle is the product of the charge q times the electric field,

$$\mathbf{F} = -\frac{iq^2}{(2\pi)^3} \int d^3k \, \frac{\mathbf{k}}{k^2 \, \epsilon(\mathbf{k} \cdot \mathbf{v})}, \qquad (4.190)$$

where $\mathbf{k} \cdot \mathbf{v}$ is identified as the angular temporal frequency ω. The quantity $1/\epsilon$ is complex, with the real part even and the imaginary part odd. The real part integrates to zero, while only the imaginary part survives. We therefore write

$$\mathbf{F} = -\frac{q^2}{(2\pi)^3} \int d^3k \, \frac{\mathbf{k}}{k^2} \, \Im \left[\frac{-1}{\epsilon(\mathbf{k} \cdot \mathbf{v})} \right]. \qquad (4.191)$$

The direction of the force is opposite to the particle velocity indicating slowing of the particle. This is evident from the axial symmetry of the problem. Assuming the particle moves in a straight line, the magnitude of the force represents the energy loss per unity path length. The integral can be evaluated in principle by resolving the wave vector $\mathbf{k}$ into axial and transverse components. In order to obtain convergence, one must subtract the vacuum contribution with no medium present. This is described in more detail by Landau and Lifshitz [56].

This represents the main result of this section. This approach has the advantage that the complex dielectric constant can be measured by light-optical means.

4.9 Small angle plural scattering of fast electrons

It is often the case where the thickness of a scattering material film exceeds the mean free path for the incident particle. A very thick bulk target can stop or reflect the incident beam. In this case, the scattering is adequately described by the diffusion equation. A commonly occurring case of considerable interest is where the scattering film is several mean free paths in thickness. This case is referred to as *plural* scattering. Diffusion has not yet set in, and it is necessary to describe the scattering in terms of a classical transport equation. This is permissible for an amorphous target, as typically the phase coherence of elastic scattering has been lost due to the random distribution of scattering centers, and the presence of inelastic processes. This section is based on earlier published work by Snyder and Scott [81], Keil, Zeitler, and Zinn [50], Crewe and Groves [21], and Groves [38].

This is distinctly different from elastic electron scattering in a crystal, where phase coherence is maintained. Here constructive interference occurs at the Bragg angles, giving rise to the familiar diffraction patterns. The following discussion does not apply to the diffraction case.

The mean free path is given by

$$\mu = \frac{1}{N\sigma},\qquad(4.192)$$

where N is the number of atoms per unit volume, and σ is the total scattering cross section. For fast electrons in a typical solid, μ for elastic scattering is proportional to the incident energy in the first Born approximation. Consequently, μ ranges from a few tens of nanometers at an incident energy of 10 KeV to a few hundreds of nanometers at 1 MeV. The sections observed in a transmission electron microscope must be thin relative to the mean free path, in order to avoid degradation of the image due to multiple scattering of the beam electrons. As this is not always possible, multiple scattering must be considered in the image formation. In this section we derive a method for understanding the scattering as a function of the sample thickness, measured in units of the mean free path.

We define a dimensionless quantity $n = z/\mu$, which we call the reduced thickness, where z is the thickness, measured in units of length. The probability P of an electron undergoing exactly j scattering events in the reduced thickness n is governed by Poisson statistics, namely

$$P_j(n) = \frac{n^j}{j!}\, e^{-n}.\qquad(4.193)$$

This can be appreciated by calculating the expectation value of the number of scattering events j for a given reduced thickness n. It is

$$\bar{j} = \sum_{j=0}^{\infty} j\, P_j(n) = n\, e^{-n} \sum_{j=1}^{\infty} \frac{n^{j-1}}{(j-1)!} = n. \tag{4.194}$$

The reduced thickness n is just the average number of scattering events. In this study we confine the discussion to small values of n, between zero and twenty, where diffusion has not yet set in.

Fast electrons incident on a thin film or bulk material undergo elastic scattering by the screened Coulomb potential of a target nucleus, and inelastic scattering by the electrons of the target material. As the nucleus is much more massive than the incident electron, classical kinematics dictates that the energy transfer is negligible, hence the designation of elastic scattering. There is appreciable momentum transfer, however. This is related to the scattering angle ϑ by (4.100). The angular distribution for elastic scattering is proportional to the differential cross section for small angles. The small angle approximation is justified for small values of n. We define a normalized angular distribution $\sigma(\theta)$ such that

$$\int_{0}^{4\pi} \sigma(\vartheta)\, d\Omega \approx 2\pi \int_{0}^{\infty} \sigma(\vartheta)\, \vartheta\, d\vartheta = 1, \tag{4.195}$$

where $d\Omega$ is the element of solid angle. Equivalently, $\sigma(\vartheta)$ is the differential elastic scattering cross section divided by the total elastic cross section in the limit of small angles $\vartheta \ll 1$. We could use the screened Coulomb scattering result (4.116),

$$\sigma(\vartheta) = \frac{\vartheta_W^2}{\pi\,(\vartheta^2 + \vartheta_W^2)^2}, \tag{4.196}$$

where ϑ_W is the screening angle, and $\sigma(\vartheta)$ is normalized to unity with respect to solid angle. In the following analysis, we will not restrict the form of $\sigma(\vartheta)$, however. In this sense the following can be regarded as completely general with respect to the detailed form of the single scattering, as long as the scattering angles are small.

We assume the elastic scattering is axially symmetric. This implies that spin polarization is unimportant, and the scattering medium is isotropic. We further assume that scattering angles associated with inelastic scattering are negligible on average, and can be ignored for the present purpose.

We will find it useful in the following to regard the scattering angle as a two-dimensional vector $\mathbf{r}'$ with components (x', y'), where $x' = dx/dz$ is the slope with respect to the transverse x–coordinate, and $y' = dy/dz$ is the slope with respect to the transverse y–coordinate. The magnitude of the scattering angle ϑ is given for small angles by

$$\vartheta \approx |\mathbf{r}'| = \sqrt{x'^2 + y'^2}. \tag{4.197}$$

Given the distribution $\sigma(\mathbf{r}')$ for single scattering, we now seek the distribution $\sigma_2(\mathbf{r}')$ for exactly two scattering events. This is

$$\sigma_2(\mathbf{r}') = \int d^2\mathbf{r}'_0 \, \sigma(|\mathbf{r}'_0|) \, \sigma(|\mathbf{r}' - \mathbf{r}'_0|) = \sigma(\mathbf{r}') * \sigma(\mathbf{r}'), \tag{4.198}$$

where $*$ denotes the two-dimensional convolution with respect to slope components. Continuing this logic, the angular distribution for exactly j scattering events is

$$\sigma_j(\mathbf{r}') = \sigma(\mathbf{r}') * \ldots * \sigma(\mathbf{r}'), \tag{4.199}$$

where the two-dimensional convolution is performed j times.

With this preparation complete, we are now in a position to state the plural scattering problem in mathematical terms: given an angular distribution $\sigma(\mathbf{r}')$ for single scattering, normalized to unity, and a mean free path μ, calculate the angular distribution $F(\mathbf{r}', z)$ for thickness z, in the presence of plural scattering. This is found by summing over all numbers of scattering events j as follows:

$$F(\mathbf{r}'; z) = \sum_{j=0}^{\infty} P_j(z/\mu) \, \sigma_j(\mathbf{r}'). \tag{4.200}$$

We now propose to eliminate the unwieldy convolution σ_j by taking the Fourier transform of both sides, and making use of the convolution theorem. The two-dimensional Fourier transform of $\sigma(\mathbf{r}')$ is defined as

$$\tilde{\sigma}(\mathbf{l}) = \int d^2\mathbf{r}'\, \sigma(\mathbf{r}')\, \exp\left[i(\mathbf{l}\cdot\mathbf{r}')\right], \qquad (4.201)$$

where $\mathbf{l}$ is the two-dimensional vector representing the transform variable conjugate to $\mathbf{r}'$. Making use of the radial symmetry, $\sigma(\mathbf{r}') = \sigma(r')$, this becomes

$$\tilde{\sigma}(\mathbf{l}) = \int_0^\infty dr'\, r'\, \sigma(r') \int_0^{2\pi} d\phi\, \exp\left(ilr'\cos\phi\right). \qquad (4.202)$$

The $\phi-$ integral can be written in terms of

$$J_0(x) = \frac{1}{2\pi}\int_0^{2\pi} d\phi\, \exp\left(i\,x\,\cos\phi\right), \qquad (4.203)$$

where J_0 is the Bessel function of zero-order. This reduces to the well-known Bessel transform,

$$\tilde{\sigma}(l) = 2\pi \int_0^\infty dr'\, r'\, J_0(lr')\, \sigma(r'), \qquad (4.204)$$

which is simply a two-dimensional Fourier transform of a radially symmetric function. Applying the same logic to F, we obtain

$$\tilde{F}(l;z) = 2\pi \int_0^\infty dr'\, r'\, J_0(lr')\, F(r';z). \qquad (4.205)$$

Taking the two-dimensional Fourier transform of both sides with respect to slope components, and making use of the convolution theorem, we obtain

$$\tilde{F}(l;z) = \sum_{j=0}^\infty P_j(n)\,[\tilde{\sigma}(l)]^j = e^{-n}\sum_{j=0}^\infty \frac{[n\tilde{\sigma}]^j}{j!}, \qquad (4.206)$$

where l is the transform variable corresponding to the scattering angle r', where $r' \ll 1$. Performing the sum, we obtain

$$\tilde{F}(l;z) = \exp\left\{-\frac{z}{\mu}[1 - \tilde{\sigma}(l)]\right\}. \qquad (4.207)$$

The solution for $F(r'; z)$ is found by performing the inverse Bessel transform,

$$F(r'; z) = \int_0^\infty dl\, l\, J_0(lr')\, \tilde{F}(l, z). \tag{4.208}$$

This integral is typically performed numerically. In doing so, one must subtract the unscattered beam $\exp(-n)$ from $\tilde{F}$, as this leads to a delta function, which is poorly behaved. This represents the solution for the angular distribution in the presence of small angle plural scattering.

It is instructive to derive $F(\mathbf{r}'; z)$ by an alternative method, which will turn out to have more general applicability. The rate of change of F with path length s can be expressed as

$$\frac{d}{ds} F(\mathbf{r}'; z) = -\frac{1}{\mu} F(\mathbf{r}'; z) + \frac{1}{\mu} \int d^2\mathbf{r}_0'\, F(\mathbf{r}_0'; z)\, \sigma(|\mathbf{r}' - \mathbf{r}_0'|). \tag{4.209}$$

The first term on the right represents scattering out of the solid angle element $d\Omega$ at $\mathbf{r}'$, while the second term on the right represents scattering into the solid angle $d\Omega$ at $\mathbf{r}'$ from all other solid angles $d\Omega_0 = d^2\mathbf{r}_0'$ at $\mathbf{r}_0'$. The quantity $1/\mu$ represents the probability per unit length that a scattering event will take place, remembering that μ is the mean free path. Using the chain rule for partial differentiation, we expand the derivative with respect to path length, obtaining

$$\frac{d}{ds} F(x', y'; z) = \left(\frac{dx'}{ds} \frac{\partial}{\partial x'} + \frac{dy'}{ds} \frac{\partial}{\partial y'} + \frac{dz}{ds} \frac{\partial}{\partial z} \right) F(x', y'; z). \tag{4.210}$$

We note that $dx'/ds = dy'/ds = 0$, as the trajectories are straight lines with constant slope between scattering events. Also, $dz/ds \approx 1$ for small angles. This leads to

$$\frac{\partial}{\partial z} F(\mathbf{r}'; z) = -\frac{1}{\mu} F(\mathbf{r}'; z) + \frac{1}{\mu} F(\mathbf{r}'; z) * \sigma(\mathbf{r}'). \tag{4.211}$$

This amounts to a transport equation, which governs the evolution of the distribution function $F(\mathbf{r}'; z)$ as the beam propogates through a thickness z.

Taking the two-dimensional Fourier transform with respect to $\mathbf{r}'$, and applying the convolution theorem, we obtain

$$\frac{\partial}{\partial z}\tilde{F}(l, z) = -\frac{1}{\mu}\tilde{F}(l, z)\cdot[1 - \tilde{\sigma}(l)], \qquad (4.212)$$

where we have made use of the radial symmetry of F and σ, as before. This is immediately integrated to reproduce the previous result.

To this point we have only considered the distribution with respect to angle or slope. It is also of considerable interest to discuss the distribution with respect to transverse coordinates. This governs the lateral broadening of electron probes in thick films, as well as the resolution in transmission electron microscopes for thick specimens. In this case we must include the two-dimensional transverse position $\mathbf{r}$ and the two-dimensional slope vector $\mathbf{r}'$. The geometry is shown in Figure 4.9. We define a distribution function $F(\mathbf{r}, \mathbf{r}'; z)$ as the probability per unit area per unit solid angle for the particle at depth z, in the presence of plural scattering. Applying the preceding logic, we expect F to satisfy

$$\frac{d}{ds}F(\mathbf{r}, \mathbf{r}'; z) = -\frac{1}{\mu}F(\mathbf{r}, \mathbf{r}'; z) + \frac{1}{\mu}\int d^2\mathbf{r}'_0\, F(\mathbf{r}, \mathbf{r}'_0; z)\,\sigma(|\mathbf{r}' - \mathbf{r}'_0|). \qquad (4.213)$$

To solve this equation for F, we begin by considering only one transverse coordinate x, and one transverse slope component x'. This is equivalent to a projection of the plural scattering problem onto the longitudinal xz-plane. The transport equation in this special case reduces to

$$\frac{d}{ds}F(x, x'; z) = -\frac{1}{\mu}F(x, x'; z) + \frac{1}{\mu}\int d^2\mathbf{r}'_0\, F(x, x'_0; z)\,\tau(x' - x'_0). \qquad (4.214)$$

The single scattering distribution $\tau(x')$ is a projection of the two-dimensional single scattering distribution $\sigma(\mathbf{r}')$. For the special case of screened Coulomb scattering, this is given by

$$\tau(x') = \int_{-\infty}^{\infty} dy'\,\sigma(x', y') = \frac{r'^2_W}{2}\frac{1}{(x'^2 + r'^2_W)^{3/2}}. \qquad (4.215)$$

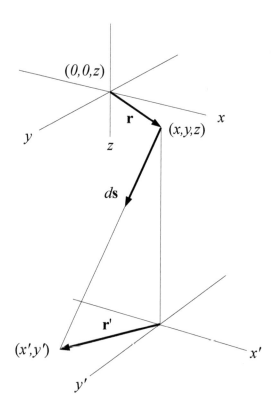

Figure 4.9: Geometry for plural scattering at depth z.

We will not assume this special dependency in the following analysis, but rather a general form for small-angle scattering $\tau(x')$.

Applying the chain rule, and expanding the total derivative as before, we obtain

$$\frac{d}{ds} F(x, x'; z) = \left(\frac{dx}{ds} \frac{\partial}{\partial x} + \frac{dx'}{ds} \frac{\partial}{\partial x'} + \frac{dz}{ds} \frac{\partial}{\partial z} \right) F(x, x'; z).$$

$$(4.216)$$

Again $dx'/ds = 0$. For small angles, $dx/ds \approx x'$ and $dz/ds \approx 1$, leading to

$$\left(x' \frac{\partial}{\partial x} + \frac{\partial}{\partial z} \right) F(x, x'; z) = -\frac{1}{\mu} F(x, x'; z) + \frac{1}{\mu} F(x, x'; z) * \tau(x').$$

$$(4.217)$$

As before, we propose to eliminate the unwieldy convolution by taking the Fourier transform of both sides, and applying the convolution theorem. We define the one-dimensional Fourier transforms as

$$
\begin{aligned}
\tilde{\tau}(l) &= \int_{-\infty}^{\infty} dx'\, \tau(x') \exp\left(ilx'\right) \\
\tilde{F}(k,l;z) &= \int_{-\infty}^{\infty} dx \int_{-\infty}^{\infty} dx'\, F(x,x';z) \exp\left[i(kx+lx')\right].
\end{aligned}
\tag{4.218}
$$

Applying the operator

$$
\int_{-\infty}^{\infty} dx \int_{-\infty}^{\infty} dx' \exp\left[i(kx+lx')\right]
\tag{4.219}
$$

to both sides from the left, and interchanging the order of integrations, we obtain the reduced equation

$$
\left(-k\frac{\partial}{\partial l}+\frac{\partial}{\partial z}\right)\tilde{F}(k,l;z) = -\frac{1}{\mu}\tilde{F}(k,l;z)\cdot[1-\tilde{\tau}(l)].
\tag{4.220}
$$

In order to integrate this equation, we propose a transformation of variables, defining the new variables

$$
\xi = l + kz, \qquad\qquad \eta = l - kz.
\tag{4.221}
$$

Applying the chain rule for partial derivatives, we find

$$
\begin{aligned}
\frac{\partial}{\partial l} &= \frac{\partial\xi}{\partial l}\frac{\partial}{\partial\xi}+\frac{\partial\eta}{\partial l}\frac{\partial}{\partial\eta} \\
\frac{\partial}{\partial z} &= \frac{\partial\xi}{\partial z}\frac{\partial}{\partial\xi}+\frac{\partial\eta}{\partial z}\frac{\partial}{\partial\eta}.
\end{aligned}
\tag{4.222}
$$

Substituting, we find

$$
-k\frac{\partial}{\partial l}+\frac{\partial}{\partial z} = -2k\frac{\partial}{\partial\eta},
\tag{4.223}
$$

and, consequently,

$$
-2k\frac{\partial}{\partial\eta}F(k,l;z) = -\frac{1}{\mu}F(k,l;z)\cdot[1-\tilde{\tau}(l)].
\tag{4.224}
$$

We have succeeded in reducing the dimensionality to a single integration variable η. This technique in the theory of partial differential equations is known as the *method of characteristics*. We can now proceed to perform the integration as

$$\ln \tilde{F}(k, l; z) = \frac{1}{2k\mu} \int_\xi d\eta \, [\, 1 - \tilde{\tau}(l) \,], \qquad (4.225)$$

where the subscript ξ signifies that ξ must be kept constant over the integration path. We treat the variable k as constant, as no derivative of k appears. We also make use of

$$2\, l = \xi + \eta, \qquad\qquad 2\, dl = d\xi + d\eta. \qquad (4.226)$$

Since $\xi = const$, and hence $d\xi = 0$ for the integration, this gives

$$\ln \tilde{F}(k, l; z) = \frac{1}{k\mu} \int_\xi^l dl \, [\, 1 - \tilde{\tau}(l) \,]. \qquad (4.227)$$

This is immediately integrated to give

$$\tilde{F}(k, l; z) = \exp \left\{ \frac{1}{k\mu} \, [\, \tilde{g}(l) - \tilde{g}(l + kz) \,] \right\} \qquad (4.228)$$

where we have defined $\tilde{g}$ by the indefinite integral

$$\tilde{g}(l) = \int dl \, [\, 1 - \tilde{\tau}(l) \,]. \qquad (4.229)$$

We note that

$$\tilde{g}(l + kz) = \tilde{g}(\xi) = const, \qquad (4.230)$$

since $\xi = const$ in the integration. The reader can verify by direct substitution that this is indeed the correct solution. It only remains to perform the inverse Fourier transform to obtain the solution, namely

$$F(x, x'; z) = \frac{1}{(2\pi)^2} \int_{-\infty}^\infty dk \int_{-\infty}^\infty dl \, \tilde{F}(l, k; z) \, \exp \, [\, -i(kx + lx') \,]. \tag{4.231}$$

Given the single scattering law $\tau(x')$, projected onto the xz-plane, we thus obtain the projected plural scattering distribution

$F(x, x'; z)$ in principle. Typically, this last integral is performed numerically, after subtracting the unscattered beam $\exp(-z/\mu) = \exp(-n)$ from $\tilde{F}$.

It is instructive to investigate several limiting cases. In the limit of zero thickness, $z = 0$, we find immediately that

$$\tilde{F}(k, l; 0) = 1. \tag{4.232}$$

Performing the inverse transform, this leads to

$$F(x, x'; 0) = \delta(x) \cdot \delta(x'), \tag{4.233}$$

thus recovering the incident beam, as required.

In the limit $k \to 0$, a Taylor expansion gives us

$$\tilde{g}(l + kz) = \tilde{g}(l) + \tilde{g}'(l)\, kz, \tag{4.234}$$

to first order in k. In this limit, $\tilde{F}$ reduces to

$$\tilde{F}(0, l; z) = \exp\left\{ -\frac{z}{\mu} [\, 1 - \tilde{\tau}(l)\,]\right\}, \tag{4.235}$$

which represents the projected angular distribution. This is expected, as $k = 0$ in Fourier space represents an integral over all x in direct space.

In the limit $l = 0$, we obtain

$$\tilde{F}(k, 0; z) = \exp\left\{ \frac{1}{k\mu} [\, \tilde{g}(0) - \tilde{g}(kz)\,]\right\}. \tag{4.236}$$

Setting $l = 0$ in Fourier space represents an integral over all scattering angles in direct space. The distribution $F(x, 0; z)$ in direct space is obtained from the inverse Fourier transform

$$F(x, 0; z) = \frac{1}{2\pi} \int_{-\infty}^{\infty} dk\, \exp(-ikx)\, \tilde{F}(k, 0; z). \tag{4.237}$$

Physically, this represents the line spread function, corresponding to scanning an incident probe beam along the infinite y-axis, and observing in the xz-plane.

With these mathematical methods established, we are now in a position to solve for the full three-dimensional distribution function $F(\mathbf{r}, \mathbf{r}'; z)$ as a function of transverse coordinate $\mathbf{r}$ and slope $\mathbf{r}'$ at depth z. The rate of change of F with path length in polar coordinates (r, ϕ) is given by the chain rule as

$$\frac{d}{ds} F(r, \phi, r', \phi'; z) = \left(\frac{dr}{ds} \frac{\partial}{\partial r} + \frac{d\phi}{ds} \frac{\partial}{\partial \phi} + \frac{dz}{ds} \frac{\partial}{\partial z} \right) F(r, \phi, r', \phi'; z),$$

(4.238)

where $dr'/ds = 0$, and $d\phi'/ds = 0$, because the trajectories form straight lines between scattering events. Making use of the axial symmetry, F is independent of azimuth ϕ, in which case the second term on the right vanishes. We note that F depends on the azimuthal slope component ϕ', as the scattering angle $\mathbf{r}'$ has a skew component in general for two or more scattering events. For small angle scattering, $dz/ds \approx 1$, in which case we can substitute

$$\frac{d}{ds} F(r, r', \phi'; z) = \left(r' \frac{\partial}{\partial r} + \frac{\partial}{\partial z} \right) F(r, r', \phi'; z).$$

(4.239)

Applying the logic of the preceding section, the transport equation is

$$\left(r' \frac{\partial}{\partial r} + \frac{\partial}{\partial z} \right) F(r, r', \phi'; z) = -\frac{1}{\mu} F(r, r', \phi'; z)$$
$$+ \frac{1}{\mu} \int d^2 \mathbf{r}'_0 \, F(r, r'_0, \phi'_0; z) \, \sigma(|\mathbf{r}' - \mathbf{r}'_0|),$$

(4.240)

where $\sigma(|\mathbf{r}'|)$ is the differential cross section for elastic scattering, normalized to unity. As before, the first term on the right represents absorption into all scattering angles from the angle of interest, and the second term on the right represents emission from all scattering angles into the angle of interest. This equation is formally similar to the preceding case. Consequently, the preceding analysis can be adopted, being mindful of the various vector components. The transformed equation is

$$\left(-k\frac{\partial}{\partial l_r} + \frac{\partial}{\partial z}\right)\tilde{F}(k, l_r, l_\phi; z) = -\frac{1}{\mu}\tilde{F}(k, l_r, l_\phi; z)\left[1 - \tilde{\sigma}(l)\right],$$

$$(4.241)$$

where (k, l_r, l_ϕ) are the Fourier transform variables corresponding to (r, r', ϕ'), respectively, and

$$l = |\mathbf{l}| = \sqrt{l_r^2 + l_\phi^2}.$$

$$(4.242)$$

Applying the method of characteristics, we define the variables

$$\xi = l_r + kz, \qquad \eta = l_r - kz, \qquad (4.243)$$

in which case, the transformed equation reduces to

$$-2k\frac{\partial}{\partial \eta}\tilde{F} = -\frac{1}{\mu}\tilde{F}\left[1 - \tilde{\sigma}(l)\right].$$

$$(4.244)$$

Integrating this with respect to η,

$$\ln\tilde{F} = \frac{1}{2k\mu}\int_\xi d\eta\left[1 - \tilde{\sigma}(l)\right].$$

$$(4.245)$$

Noting that

$$2\,l_r = \xi + \eta, \qquad 2\,dl_r = d\xi + d\eta \qquad (4.246)$$

with $\xi = const$, and hence $d\xi = 0$ for the integration. This gives

$$\ln\tilde{F} = \frac{1}{\mu k}\int_{l_r+kz}^{l_r} dl_r\left[1 - \tilde{\sigma}\left(\sqrt{l_r^2 + l_\phi^2}\right)\right].$$

$$(4.247)$$

This is immediately integrated to give

$$\tilde{F}(k, l; z) = \exp\left\{\frac{1}{k\mu}\left[\tilde{g}(l_r, l_\phi) - \tilde{g}(l_r + kz, l_\phi)\right]\right\}, \qquad (4.248)$$

where we have defined $\tilde{g}(l_r, l_\phi)$ by the indefinite integral

$$\tilde{g}(l_r, l_\phi) = \int dl_r\left[1 - \tilde{\sigma}\left(\sqrt{l_r^2 + l_\phi^2}\right)\right], \qquad (4.249)$$

and l_ϕ is regarded as constant under the integral.

Investigating the limiting cases, we see that, for $z = 0$,

$$\tilde{F}(k, l_r, l_\phi; 0) = 1. \qquad (4.250)$$

Performing the inverse transform,

$$F(r, r', \phi'; 0) = \delta(r)\,\delta(\mathbf{r}'), \qquad (4.251)$$

thus recovering the incident beam, as required.

In the limit $k \to 0$, we can write the Taylor expansion for $\tilde{g}(l_r + kz, l_\phi)$ to first order as

$$\tilde{g}(l_r + kz, l_\phi) = \tilde{g}(l_r, l_\phi) + kz \cdot \frac{\partial}{\partial l_r}\,\tilde{g}(l_r, l_\phi), \qquad (4.252)$$

in which case,

$$\tilde{F}(k, l_r, l_\phi; z) = \tilde{F}(0, l_r, l_\phi; z) = \exp\left\{-\frac{z}{\mu}\left[1 - \tilde{\sigma}(l)\right]\right\}, \qquad (4.253)$$

remembering that $l = |\mathbf{l}| = \sqrt{l_r^2 + l_\phi^2}$. This is immediately recognizable as the angular distribution. This is expected, as $k = 0$ in Fourier space represents an integral over the entire range of radial coordinate, $0 \le r < \infty$ in direct space. This result is superfluous, as it was derived previously by simpler methods.

Finally, setting $\mathbf{l} = 0$, we find

$$\tilde{F}(k, l_r, l_\phi; z) = \tilde{F}(k, 0, 0; z) = \exp\left\{\frac{1}{k\mu}\left[\tilde{g}(0, 0) - \tilde{g}(kz, 0)\right]\right\}, \qquad (4.254)$$

where this represents the integral over all slopes $|\mathbf{l}|$ in direct space. The distribution in the transverse radial coordinate r is found by performing the inverse Bessel transform,

$$F(r; z) = \int_0^\infty dk\, k\, J_0(kr)\, \tilde{F}(k, 0, 0; z). \qquad (4.255)$$

This represents the radial point spread function at depth z, in the presence of plural scattering. This being the case, it follows that $\tilde{F}(k, 0, 0; z)$ represents the modulation transfer function, as this is the Fourier transform of the point spread function. This completes the general solution to the plural scattering problem.

Chapter 5

Electron emission from solids

Every practical electron beam instrument relies on a stable, long-lived electron source. The most commonly used sources extract electrons from a bulk metal or semiconductor, and accelerate the particles across a vacuum gap using the electric field of an electrode at a positive potential relative to the source. The beam thus appears to originate from an apparent source, real or virtual, seen looking back from the electron optical system.

For practical purposes this apparent source is characterized by its measurable macroscopic properties. These include beam energy, current, lateral intensity distribution, angular intensity distribution, and energy spread. These quantities are a function of the physical source properties, including geometry and material. They also depend on the temperature and applied electric field as controllable operating parameters. Because of brightness conservation, the macroscopic source properties govern the optical properties of the entire optical system.

Electron emission is a fundamentally quantum mechanical process. A great deal of insight can be gained by regarding the bulk emitter material in terms of a relatively simple model originally proposed by Sommerfeld [82]. In this model, the conduction electrons are approximately free to diffuse throughout the bulk material. Electrons in the conduction band occupy energy states. The Pauli exclusion principle dictates that no more than one electron can occupy any given state. The average occupation number for a state obeys Fermi–Dirac statistics, and is between zero and one. In the limit where the absolute temperature T approaches zero, all states with energy ε in the range $0 \leq \varepsilon \leq \zeta$ are occupied by one electron, where ζ is called the Fermi energy. In this limit all states with energy higher than the Fermi energy are unoccupied.

In the following we assume that the emission surface is planar, and infinite in lateral extent. In this approximation the problem can be regarded as spatially one-dimensional, with the x-axis perpendicular to the emission surface. Some fraction of the conduction electrons drift to the surface, where they can be emitted into the vacuum to form a beam. Once emitted, an electron experiences a Coulomb force which tends to attract it back toward the emission surface. This is called an *image* force, and is described in detail in the following section. It gives rise to a potential energy barrier which must be overcome in order for the electron to be emitted into the vacuum.

In this one-dimensional model we consider the potential energy of an electron to be zero everywhere inside the bulk material. Electrons are free to drift throughout the bulk material, with a net flux incident on the emission surface from within the material. For electrons with a specific total energy W within the bulk material, we assume a current density $J(W)$ incident on the emission surface from within. Here $J(W)$ has dimensions of charge per unit transverse area per unit time per unit energy W. We further assume a single electron with energy W has a probability $D(W)$ of overcoming the potential barrier, to be emitted into the vacuum. The total emission current density j is then given by

$$j = \int_0^\infty dW \, J(W) \, D(W), \qquad (5.1)$$

where we have integrated over all possible values of the energy W.

In practice many sources are not planar. Furthermore, the energy bands of the solid emitter material tend to bend at the interface with the vacuum. This band-bending can be enhanced by preparing the emitter surface with an additional surface layer. These factors affect the actual emission, and the resulting macroscopic source properties. We define the *electron affinity* as the energy needed in practice to remove a single electron from the emitter. Because of these factors, the following analysis must be regarded as approximate. As is often the case, one replaces an intractable problem by a related problem which can be solved, producing an approximate result.

To summarize, the emission current density depends on the temperature and the applied electric field. In the limit of zero applied electric field F and elevated temperature T, the current is called *thermionic* emission. In the limit of zero temperature T and elevated electric field F, the current is called *field* emission. The central problem of this chapter is to determine the emitted current density j as a general function of temperature T and applied electric field F.

5.1   The image force

An electron which has been emitted into the vacuum experiences an electrostatic force which attracts it back toward the emission surface. This can be understood by examining the Coulomb force on an electron with charge $-e$ located in the vacuum a distance x from a planar conductive surface at ground potential. This situation is completely equivalent to a configuration where the planar surface is replaced by a charge $+e$ located behind the emission surface at a distance $2x$ from the electron. This fictitious charge

$+e$ is called an *image* charge, and the attractive force is called the image force. This was first understood by Nordheim [66]. The magnitude of the force $\mathcal{F}$ on the electron is given by

$$\mathcal{F}(x) = \frac{-e^2}{16\pi\epsilon_0\,x^2}, \tag{5.2}$$

where the minus sign indicates that the vector Coulomb force points in the negative x-direction, back toward the emission surface.

We consider a virtual displacement of the electron from coordinate $+x$ to $+\infty$ in the presence of the force $F(x)$. This results in a change in potential energy $U(x)$ given by

$$U(x) = \int_x^\infty \mathcal{F}(\xi)\,d\xi = \frac{-e^2}{16\pi\epsilon_0\,x}, \tag{5.3}$$

where this is the work needed to remove the electron from $+x$ to $+\infty$. This properly accounts for the fact that the image charge undergoes a virtual displacement equal and opposite to the electron. An individual electron must have enough energy to surmount the potential energy barrier in order to be emitted into the vacuum. Alternatively the electron can tunnel through the barrier in the presence of an applied electric field. In either case, the expression (5.1) for j applies.

5.2 The incident current density

We now turn our attention to the current density $J(W)$ of electrons with total energy W incident on the emission surface from within the bulk material in one spatial dimension. Within a metal the conduction electrons are approximately free. We can therefore choose the potential energy to be zero. We denote the total energy of a single electron inside the metal in three spatial dimensions by ε. In the following we deduce the one-dimensional properties from the three-dimensional properties, making use of the planar symmetry.

The density of energy states for a nearly free electron within the solid is given in three dimensions (3.73) as

$$\frac{dN}{d\varepsilon} = \frac{8\pi V}{h^3}\sqrt{2m^3\varepsilon}, \tag{5.4}$$

where V is the volume, h is Planck's constant, and m is the electron mass. This is the number of available states per unit energy interval $d\varepsilon$. Each electron energy level has two spin states with the same total energy ε. To properly account for this, we have multiplied the right-hand side of (3.73) by two. The Pauli exclusion principle permits, at most, one electron occupying a given state.

The expectation value of the occupation number of a state of energy ε is governed by Fermi-Dirac statistics, and is given by

$$n(\varepsilon) = \left[\exp\left(\frac{\varepsilon - \zeta}{kT}\right) + 1\right]^{-1}, \tag{5.5}$$

where k is Boltzmann's constant, and T is the absolute temperature. The energy ζ is commonly referred to as the chemical potential per atom, and alternatively as the Fermi energy. It is easy to verify that $0 \le n(\varepsilon) \le 1$, consistent with the Pauli exclusion principle. It follows that the average charge density within the material is given as a function of total energy ε by

$$\begin{aligned} \rho(\varepsilon) &= \frac{e}{V}\left(\frac{dN}{d\varepsilon}\right)n(\varepsilon) \\ &= \frac{8\pi e}{h^3}\sqrt{2m^3\varepsilon}\,n(\varepsilon), \end{aligned} \tag{5.6}$$

where $\rho(\varepsilon)\,d\varepsilon$ is the charge per unit volume of conduction electrons with total energy between ε and $\varepsilon + d\varepsilon$. Assuming the potential energy is zero everywhere, the total energy ε is given in terms of the electron velocity v by

$$\varepsilon = \tfrac{1}{2}mv^2. \tag{5.7}$$

Since the energy ε depends only on the magnitude of the velocity and not on the direction, it follows that the electron velocities are distributed isotropically with respect to propagation direction within the bulk material in this approximation.

Choosing the x-axis to be perpendicular to the emission surface, the current density component $j_x(\varepsilon)$ is given by

$$
\begin{aligned}
j_x(\varepsilon) \; &= \; \rho(\varepsilon)\, v_x \\
&= \; \rho(\varepsilon)\, \sqrt{\frac{2\varepsilon}{m}}\, \cos\theta, \tag{5.8}
\end{aligned}
$$

where v_x is the x-component of the velocity and θ is the polar angle which the velocity vector makes with the x-axis. From (5.5, 5.6, 5.8) we obtain the x-component of the current density as

$$
j_x(\varepsilon) = \frac{16\pi m e}{h^3}\, \varepsilon\, n(\varepsilon)\, \cos\theta \tag{5.9}
$$

for $0 \le \theta \le \pi$. Next we define the differential

$$
dj_x(\varepsilon) = j_x(\varepsilon)\, \frac{d\Omega}{4\pi}, \tag{5.10}
$$

where $d\Omega$ is the solid angle element given by $d\Omega = 2\pi \sin\theta\, d\theta$. Substituting,

$$
dj_x(\varepsilon) = \frac{8\pi m e}{h^3}\, \varepsilon\, n(\varepsilon)\, \cos\theta \, \sin\theta\, d\theta. \tag{5.11}
$$

At this point we define the total energy W in the x-direction as

$$
W = \tfrac{1}{2} m v_x^2 = \varepsilon \cos^2\theta. \tag{5.12}
$$

Substituting,

$$
dj_x(\varepsilon) = \frac{8\pi m e}{h^3}\, W\, n(W\, \sec^2\theta)\, \tan\theta\, d\theta. \tag{5.13}
$$

We define the current density $J(W)$ in the one-dimensional problem according to

$$
dJ(W)\, dW = dj_x(\varepsilon)\, d\varepsilon, \tag{5.14}
$$

where this ensures that the total current is the same for the one-dimensional and three-dimensional problems. Substituting,

$$dJ(W) = \frac{8\pi me}{h^3} W \, n(W \sec^2 \theta) \sec^2 \theta \tan \theta \, d\theta. \tag{5.15}$$

We define a new variable ξ by

$$\xi = \sec^2 \theta, \qquad d\xi = 2 \sec^2 \theta \tan \theta \, d\theta. \tag{5.16}$$

Substituting, making use of (5.5), and integrating over the range $1 \leq \xi < \infty$, we find

$$J(W) = \frac{4\pi meW}{h^3} \int_1^\infty d\xi \left[\exp\left(\frac{W\xi - \zeta}{kT}\right) + 1 \right]^{-1}. \tag{5.17}$$

Performing the integral is straightforward, and is left as an exercise for the reader. We obtain the result

$$J(W) = \frac{4\pi mekT}{h^3} \ln \left[\exp\left(\frac{\zeta - W}{kT}\right) + 1 \right], \tag{5.18}$$

where $J(W)$ has dimensions of current per unit transverse area per unit energy interval. This is the main result of this section. It is identical with the result obtained by Kemble [52], and used later by Murphy and Good [64].

Anticipating the case of cold field emission, it is useful to explore the limit $T \to 0$. We obtain

$$J(W) \approx \frac{4\pi me}{h^3} (\zeta - W), \tag{5.19}$$

where we note that $0 < W \leq \zeta$ in this limit.

Using the general expression (5.18) for the incident current density $J(W)$ as a function of the total energy W in one spatial dimension, we next proceed to calculate the transmission probability $D(W)$, and the resulting emission current density j for various combinations of the temperature T and the applied field F. This is the topic of the following sections.

Problems

1. Perform the integral (5.17) to obtain the result (5.18).

2. Verify the result (5.19) for the limit $T \to 0$ by repeating the procedure of this section, making use of the fact that

$$
n(\varepsilon) = \left[\exp\left(\frac{\varepsilon - \zeta}{kT} \right) + 1 \right]^{-1} = \begin{cases} 1, & 0 \le \varepsilon \le \zeta \\ 0, & \varepsilon > \zeta. \end{cases} \tag{5.20}
$$

3. Derive an analytical expression for the Fermi energy ζ of a metal based on the number of conduction band electrons per unit volume and the density of states with respect to total energy ε.

5.3   Thermionic emission

In the case of zero applied electric field and elevated temperature, the energy needed for an electron to surmount the potential barrier is thermal. This is called *thermionic* emission. The central problem in this section is to calculate the emission current density j for thermionic emission. We make use of (5.1, 5.18). The task remains to calculate the probability $D(W)$ that an electron with total energy W will be transmitted across the barrier.

The potential energy $U(x)$ associated with the image force is given by (5.3). This is plotted as a function of coordinate x in the direction normal to the emission surface in Figure 5.1. The surface of the metal is at coordinate $x = 0$. The left region $x < 0$ represents the interior of the bulk emitting material, and the right region

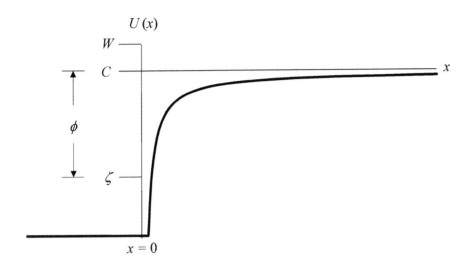

Figure 5.1: Energy diagram for thermionic emission.

$x > 0$ represents the vacuum. The solid curve is the potential energy $U(x)$.

We consider a single electron with energy W inside the bulk emitter material with $x \leq 0$. We approximate the potential energy $U(x)$ by a square barrier. The problem of scattering of matter waves by a square barrier in one dimension is treated in many books on elementary quantum mechanics, see for example Liboff [59].

Inside the bulk material with $x \leq 0$ the wave function is a superposition of a right-propagating incident wave plus a left-propagating reflected wave. This is represented by

$$u(x \leq 0) = a_+ \, e^{+ik_1 x} + a_- \, e^{-ik_1 x}, \tag{5.21}$$

where the complex constants a_+ and a_- have yet to be determined. The direction of propagation can be verified by forming the complete wave function $\psi(x, t)$ given by

$$\psi(x, t) = u(x) \, e^{iWt/\hbar}, \tag{5.22}$$

which represents two counterpropagating, travelling waves. The wave number k_1 is given by

$$k_1 = \sqrt{\frac{2m}{\hbar^2} W}, \tag{5.23}$$

where we intentionally choose the positive root. In the vacuum with $x \geq 0$ the wave function is

$$u(x \geq 0) = b_+ \, e^{+ik_2 x}, \tag{5.24}$$

where the complex constant b_+ has yet to be determined. The wave number k_2 is given by

$$k_2 = \sqrt{\frac{2m}{\hbar^2} (W - C)}, \tag{5.25}$$

We assume that no left-propagating wave exists in the vacuum region $x \geq 0$. We need only consider energy $W \geq C$, since there can be no transmission for $W < C$.

We require that the wave function and its first derivative be continuous at $x = 0$. This leads to the coupled equations

$$
\begin{aligned}
a_+ + a_- &= b_+ \\
a_+ - a_- &= \frac{k_2}{k_1} b_+.
\end{aligned}
\tag{5.26}
$$

These can be immediately reduced to

$$
\begin{aligned}
\frac{b_+}{a_+} &= \frac{2}{1 + k_2/k_1} \\
\frac{a_-}{a_+} &= \frac{1}{2} \cdot \frac{1 - k_2/k_1}{1 + k_2/k_1}.
\end{aligned}
\tag{5.27}
$$

Each propagating wave has an asssociated probability current given by

$$j = \frac{i\hbar}{2m} \left[u(x)\, \bar{u}'(x) - \bar{u}(x)\, u'(x) \right]. \tag{5.28}$$

Substituting the wave functions, we have

$$j_+(x \leq 0) = \frac{\hbar k_1}{m} |a_+|^2$$

$$j_-(x \leq 0) = -\frac{\hbar k_1}{m} |a_-|^2$$

$$j_+(x \geq 0) = \frac{\hbar k_2}{m} |b_+|^2. \tag{5.29}$$

These represent the incident, reflected, and transmitted currents, respectively. The transmission probability $D(W)$ is the ratio of the transmitted current divided by the incident current. This is

$$D(W) = \frac{4\sqrt{1 - C/W}}{\left(1 + \sqrt{1 - C/W}\right)^2}, \tag{5.30}$$

where we have made use of

$$\frac{k_2}{k_1} = \sqrt{\frac{W - C}{W}}, \tag{5.31}$$

where $0 \leq k_2/k_1 < 1$ for $C \leq W < \infty$. Also, $D(W) = 0$ for $W \leq C$, since it is impossible for an electron to surmount or tunnel through the potential barrier in this case. Also, $0 \leq D(W) \leq 1$, as required for a probability.

We are now in a position to calculate the emission current density j, given by (5.1). Making use of (5.18, 5.30), we have

$$j = \frac{4\pi m e k T}{h^3} \int_C^\infty dW \ln\left[\exp\left(\frac{\varsigma - W}{kT}\right) + 1\right]$$

$$\cdot \frac{4\sqrt{1 - C/W}}{\left(1 + \sqrt{1 - C/W}\right)^2}. \tag{5.32}$$

This integral cannot easily be evaluated as a closed-form expression, but is amenable to straightforward numerical evaluation.

Considerable physical insight can be gained by approximating

$D(W) \approx 1$, corresponding to perfect transmission for energies $W \geq C$. We define a new variable $\xi = W - C$. We further define a quantity ϕ called the *work function* by

$$\phi = C - \zeta. \tag{5.33}$$

Physically, ϕ represents the energy which must be supplied to an electron at the Fermi level in order for emission to occur. This quantity is a unique property of the bulk material. In practice, the work function differs from the electron affinity defined above for an actual emitter.

The integral for the emission current density j is approximated as

$$j = \frac{4\pi mekT}{h^3} \int_0^\infty d\xi \, \ln \left[\exp \left(\frac{-\phi - \xi}{kT} \right) + 1 \right]. \tag{5.34}$$

We further approximate

$$\exp \left(\frac{-\phi - \xi}{kT} \right) \ll 1. \tag{5.35}$$

Expanding the logarithm, the integral for j becomes

$$j = \frac{4\pi mekT}{h^3} \exp \left(-\frac{\phi}{kT} \right) \int_0^\infty d\xi \, \exp \left(-\frac{\xi}{kT} \right). \tag{5.36}$$

The integral over ξ is readily performed. The emission current density j is given by

$$j = \frac{4\pi me}{h^3} (kT)^2 \exp \left(-\frac{\phi}{kT} \right). \tag{5.37}$$

This represents the main result of this section. It is known as the Richardson–Dushman law for thermionic emission [60].

5.4 Field emission

Electrons can be extracted from the surface of a bulk metal by applying an electric field. This process is known as *field emission* or *cold emission*. It relies on the quantum-mechanical process of a single electron *tunneling* through a potential barrier. This process takes place even at very low temperature. Our goal is to obtain an explicit expression for the emission current density j in terms of the applied field F, and the work function ϕ, in the limit of low temperature T. This was first understood theoretically in 1927 by Fowler and Nordheim [31]. This source is now widely used in a variety of practical instruments. It is distinguished by its exceptionally high brightness. In the following sections we proceed to derive the current density from first principles of quantum mechanics.

For the present purpose we can consider a conduction electron to be free inside the bulk metal at $x \leq 0$. The potential energy is thus given here by $U(x) = 0$. Outside the metal at $x \geq 0$, the potential energy is given by

$$U(x) = C - Fx - \frac{e^2}{16\pi\epsilon_0\, x}, \qquad (5.38)$$

where the first term on the right is a constant, associated with the energy in the vacuum. The second term on the right is the potential energy associated with the applied electric field, where F is given by the product of the electron charge e times the electric field in volts per meter. In this notation F has units of force, which is equivalent to energy per unit distance. The third term on the right in (5.38) is the potential energy associated with the image force. This form for the potential energy was first understood by Nordheim [66].

For now we will ignore the third term, since it is relatively weak in the limit of high electric field. This approximation was used by Fowler and Nordheim [31]. The potential energy is plotted in this approximation as a function of x in Figure 5.2. In a later section

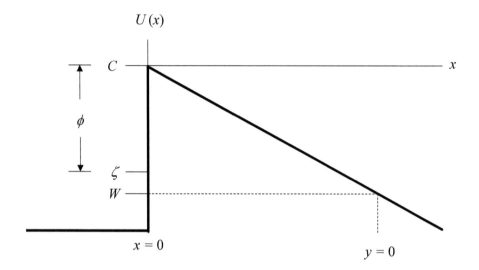

Figure 5.2: Approximate energy diagram for cold field emission.

we will include this image force term to improve the accuracy and generality of the calculation.

In the limit of zero absolute temperature T, conduction electrons within the bulk material occupy all energy states up to the maximum energy ζ, which is the Fermi energy. No states above this energy are occupied. Classically no emission can occur, because the topmost filled energy level is below the top of the potential energy barrier. Quantum mechanically a conduction electron incident on the barrier from within the buk can tunnel through the barrier, and be emitted into the vacuum.

We designate $D(W)$ as the probability that an electron with energy W inside the material will tunnel through the potential barrier. Again we designate the incident current density per unit energy from within the material as $J(W)$. We seek the emission current density, which is the incident current density $J(W)$ times the transmission probability $D(W)$, integrated over all possible energies W, where $0 \le W \le \zeta$. This is

$$j = \int_0^\zeta dW\, J(W)\, D(W), \qquad (5.39)$$

where ζ is the Fermi energy.

To find the field emission current density j, we must first solve for the transmission probability $D(W)$. All relevant information is contained in the stationary state spatial wave function $u(x)$. This is found by solving Schrödinger's equation separately inside and outside the bulk metal for $u(x)$, and matching the two solutions $u(x)$ and their first derivatives $u'(x)$ at the emission surface $x = 0$.

Considering first the region interior to the bulk metal at $x \le 0$, only the energy in the x-direction is relevant, as the transverse component has no effect. Schrödinger's equation is given from the preceding analysis in Chapter 2 as

$$\left(\frac{d^2}{dx^2} + k^2 \right) u(x) = 0. \qquad (5.40)$$

According to the earlier analysis, the wave vector k is related to the scalar kinetic momentum p by

$$p = \hbar k = \pm\sqrt{2mW}, \qquad (5.41)$$

where W is the total energy of a single electron in the x-direction. The state function is the sum of two linearly independent solutions $u_\pm(x)$ for the eigenfunction $u(x)$,

$$\begin{aligned} u_+(x) &= a_+ e^{+ikx} \\ u_-(x) &= a_- e^{-ikx}, \end{aligned} \qquad (5.42)$$

where $a_\pm$ represent two arbitrary complex constants. One can immediately verify the solutions $u_\pm(x)$ by direct substitution into the differential equation.

The total energy eigenvalue W is the same for both solutions, and is given by

$$W = \frac{\hbar^2 k^2}{2m}$$

$$= \tfrac{1}{2} m v_x^2. \tag{5.43}$$

For now we consider a single, specific energy W with associated wave vector k.

The probability current is

$$j(x) = \frac{i\hbar}{2m} \left[u(x)\,\bar{u}'(x) - \bar{u}(x)\,u'(x) \right], \tag{5.44}$$

where a bar over a quantity indicates complex conjugation. Substituting $u_{\pm}(x)$ above, we identify right- and left-propagating currents

$$j_+(x \leq 0) = +\frac{\hbar k}{m} |a_+|^2$$

$$j_-(x \leq 0) = -\frac{\hbar k}{m} |a_-|^2, \tag{5.45}$$

respectively, where j_+ is the current incident on the potential barrier from left to right inside the bulk at $x \leq 0$, and j_- is the current reflected by the barrier from right to left. The algebraic sum of these is the tunneling current transmitted through the barrier.

For $x \geq 0$, the vacuum, Schrödinger's equation is

$$\left[-\frac{\hbar^2}{2m}\frac{d^2}{dx^2} + U(x) \right] u(x) = W\,u(x), \tag{5.46}$$

where the potential energy $U(x)$ is given approximately by

$$U(x) \approx C - F\,x. \tag{5.47}$$

The quantity F is the electron charge e times the applied electric field, and has the dimension of force in nt. Equivalently we write this as

$$\frac{d^2}{dx^2} u(x) + \frac{2m}{\hbar^2} [W - C + Fx] u(x) = 0. \tag{5.48}$$

We rewrite this as

$$\frac{d^2}{dx^2} u(x) + \alpha^3 (x - \beta) u(x) = 0, \tag{5.49}$$

where we have defined the constants

$$\alpha^3 = \frac{2mF}{\hbar^2}$$

$$\beta = \frac{C - W}{F}. \tag{5.50}$$

We define a new variable $y(x)$ as

$$y(x) \equiv \alpha (\beta - x). \tag{5.51}$$

The differential equation for $u(x)$ is thus transformed into

$$\left(\frac{d^2}{dy^2} - y \right) Y(y) = 0, \tag{5.52}$$

where we have defined the eigenfunction $Y(y)$ according to

$$Y(y) \equiv u[x(y)]. \tag{5.53}$$

Two linearly independent solutions for $Y(y)$ exist, and are designated

$$Y(y) = \begin{cases} Ai(y) \\ Bi(y). \end{cases} \tag{5.54}$$

The functions Ai and Bi are called Airy functions. Their properties are well-known [1]. The vacuum is represented by large positive values of x, corresponding to large negative values of y. The Airy functions have asymptotic forms for $y \ll 0$ given by

$$Ai(y) \approx \frac{1}{\sqrt{\pi} (-y)^{1/4}} \sin \left[\tfrac{2}{3} (-y)^{3/2} + \tfrac{\pi}{4} \right]$$

$$Bi(y) \approx \frac{1}{\sqrt{\pi} (-y)^{1/4}} \cos \left[\tfrac{2}{3} (-y)^{3/2} + \tfrac{\pi}{4} \right]. \tag{5.55}$$

We form the linear combination

$$Y(y) = b_+ \left[Bi(y) + iAi(y) \right], \tag{5.56}$$

defined for all y, where b_+ is an arbitrary complex constant. For y large and negative, $Y(y)$ has the asymptotic form

$$Y(y) \approx b_+ \frac{1}{\sqrt{\pi}\,(-y)^{1/4}} \exp\left\{ i\left[\tfrac{2}{3}(-y)^{3/2} + \tfrac{\pi}{4} \right] \right\}. \tag{5.57}$$

This represents a wave which propagates to the right in the coordinate x in the vacuum. This is a necessary condition, since we must assume no left-propagating wave can exist in the vacuum. We therefore adopt this form as our solution $Y(y)$ for all y. We further define the probability current $J(y)$ as

$$J(y) = J[\,y(x)\,] = j(x), \tag{5.58}$$

where $j(x)$ is the current defined above. We notice from (5.44) that the wave function must have an imaginary part in order to have nonzero probability current. The wave function $Y(y)$ satisfies this requirement. Substituting, we obtain

$$J(y) = -\frac{i\hbar\alpha}{2m} \left[Y(y)\,\bar{Y}'(y) - \bar{Y}(y)\,Y'(y) \right]. \tag{5.59}$$

The first derivative $Y'(y)$ is given by

$$Y'(y) = b_+ \left[Bi'(y) + iAi'(y) \right]. \tag{5.60}$$

It is straightforward to evaluate the current $J(y)$, noticing that $Ai(y)$ and $Bi(y)$ are real-valued for y real. After some algebra we obtain

$$J(y) = \frac{\hbar\alpha}{2m}\, |b_+|^2 \left[Ai(y)\,Bi'(y) - Ai'(y)\,Bi(y) \right]. \tag{5.61}$$

The quantity in square brackets is the conserved Wronskian, and has the value π^{-1}. The current $J(y)$ reduces to

$$J(y) = \frac{\hbar\alpha}{\pi m}\, |b_+|^2. \tag{5.62}$$

This is independent of coordinate, as required by the fact that it is proportional to the conserved Wronskian. We identify

$$j_+(x \geq 0) = J[y(x)] = \frac{\hbar\alpha}{\pi m}|b_+|^2 \qquad (5.63)$$

as the tunneling current propagating from left to right for $x \geq 0$. This result will prove useful later.

Next we must match the solutions and their derivatives at $x = 0$, which is the emission surface. For $x \leq 0$ inside the bulk material we form the solution and its first derivative as

$$\begin{aligned} u(x) &= a_+ e^{+ikx} + a_- e^{-ikx} \\ u'(x) &= ik\,[\,a_+ e^{+ikx} - a_- e^{-ikx}\,]. \end{aligned} \qquad (5.64)$$

For $x \geq 0$ in the vacuum we form

$$\begin{aligned} u(x) &= Y[y(x)] \\ u'(x) &= -\alpha\,Y'(y). \end{aligned} \qquad (5.65)$$

Matching the solutions and first derivatives at $x = 0$, equivalently $y = \alpha\beta$, we have two simultaneous equations,

$$\begin{aligned} a_+ + a_- &= Y(\alpha\beta) \\ a_+ - a_- &= \frac{i\alpha}{k} Y'(\alpha\beta). \end{aligned} \qquad (5.66)$$

Solving for a_+ and a_- we find

$$\begin{aligned} a_+ &= \tfrac{1}{2}Y(\alpha\beta) + \frac{i\alpha}{2k} Y'(\alpha\beta) \\ a_- &= \tfrac{1}{2}Y(\alpha\beta) - \frac{i\alpha}{2k} Y'(\alpha\beta). \end{aligned} \qquad (5.67)$$

Substituting for Y and Y' above, we find

$$\begin{aligned} a_+ &= b_+ \left\{ \tfrac{1}{2}[\,Bi(\alpha\beta) + iAi(\alpha\beta)\,] + \frac{i\alpha}{2k}\,[\,Bi'(\alpha\beta) + iAi'(\alpha\beta)\,] \right\} \\ a_- &= b_+ \left\{ \tfrac{1}{2}[\,Bi(\alpha\beta) + iAi(\alpha\beta)\,] - \frac{i\alpha}{2k}\,[\,Bi'(\alpha\beta) + iAi'(\alpha\beta)\,] \right\}. \end{aligned}$$

$$(5.68)$$

The fraction of electrons incident on the barrier from within the bulk, which tunnel through the barrier to be emitted into the vacuum is given by

$$D(W) = \frac{j_+(x \geq 0)}{j_+(x \leq 0)},$$ (5.69)

where this depends on the energy W. Substituting the above expressions for the two currents, this is

$$D(W) = \frac{\alpha}{\pi k} \frac{|b_+|^2}{|a_+|^2}.$$ (5.70)

Evaluating the absolute square of the coefficients, this becomes

$$D(W) = \frac{4\alpha}{\pi k} \cdot$$
$$\left\{ Ai^2(\alpha\beta) + Bi^2(\alpha\beta) + \frac{2\alpha}{\pi k} + \frac{\alpha^2}{k^2} \left[Ai'^2(\alpha\beta) + Bi'^2(\alpha\beta) \right] \right\}^{-1}$$

(5.71)

where we have again made use of the conserved Wronskian. Substituting from above,

$$\alpha\beta = \left(\frac{2m}{\hbar^2}\right)^{1/3} F^{-2/3} (C - W)$$
$$\frac{\alpha}{k} = \left(\frac{2m}{\hbar^2}\right)^{-1/6} F^{1/3} W^{-1/2}.$$ (5.72)

It is left as an exercise for the reader, see Problems below, to substitute some reasonable values. This represents a formal solution for the tunneling probability $D(W)$. It is possible in principle to evaluate this numerically using the known series expansions for the Airy functions and their derivatives [1].

Additional physical insight can be gained by approximating these quantities. To this end we invoke the asymptotic forms for $y \gg 0$,

$$Ai(y) \approx \frac{1}{2\sqrt{\pi} \, y^{1/4}} \exp\left(-\frac{2}{3} y^{3/2}\right)$$

$$Ai'(y) \approx -\frac{y^{1/4}}{2\sqrt{\pi}} \exp\left(-\tfrac{2}{3} y^{3/2}\right)$$

$$Bi(y) \approx \frac{1}{\sqrt{\pi}\, y^{1/4}} \exp\left(\tfrac{2}{3} y^{3/2}\right)$$

$$Bi'(y) \approx \frac{y^{1/4}}{\sqrt{\pi}} \exp\left(\tfrac{2}{3} y^{3/2}\right), \tag{5.73}$$

where we set $y = \alpha\beta$ at the interface between the bulk and the vacuum. We implicitly assume that $y \gg 0$. Substituting these asmptotic forms into the above expression for $D(W)$, we see that the terms in Bi and Bi' dominate. Retaining only these terms, we obtain the approximation

$$D(W) = \frac{4\alpha}{k} \left(\frac{1}{\sqrt{\alpha\beta}} + \frac{\alpha^2 \sqrt{\alpha\beta}}{k^2}\right)^{-1} \exp\left[-\tfrac{4}{3} (\alpha\beta)^{3/2}\right]. \tag{5.74}$$

Substituting for $\alpha\beta$ and α/k above, this leads immediately to an expression for the tunneling probability,

$$D(W) = \frac{4\sqrt{W(C-W)}}{C} \exp\left[-\frac{4}{3F}\left(\frac{2m}{\hbar^2}\right)^{1/2} (C-W)^{3/2}\right]. \tag{5.75}$$

This is the probability that an electron with energy W will tunnel through the barrier. This represents one of the main results of this section. We are now in a position to calculate the emission current density j based on (5.1, 5.19, 5.75). The field emission current density j is given from (5.1) as

$$j = \int_0^\zeta dW\, D(W)\, J(W), \tag{5.76}$$

where only states with $0 \leq W \leq \zeta$ are occupied in the limit $T \to 0$. Substituting for $D(W)$ and $J(W)$ this becomes

$$j = \frac{16\pi em}{h^3 C} \int_0^\zeta dW\, \sqrt{W(C-W)}\, (\zeta - W)$$

$$\exp\left[-\frac{4}{3F}\sqrt{\frac{2m}{\hbar^2}} (C-W)^{3/2}\right]. \tag{5.77}$$

We define a variable ξ as

$$\xi = (C - W)^{3/2}. \tag{5.78}$$

The integral (5.76) now takes the form

$$j = \int d\xi \, U(\xi) \, e^{-a\xi}, \tag{5.79}$$

where $U(\xi)$ is not to be confused with the potential energy above. This integral can be performed in principle, since the integrand is well-behaved over the range of integration [31]. A useful approximation can be obtained from the series representation

$$\int d\xi \, U(\xi) \, e^{-a\xi} = -\frac{1}{a} e^{-a\xi} \left[U(\xi) + \frac{U'(\xi)}{a} + \frac{U''(\xi)}{a^2} + \cdots \right], \tag{5.80}$$

which the reader can immediately verify by differentiating both sides with respect to ξ. The first term in the series vanishes, because the function U vanishes at the two end points. The second and successive terms are infinite, owing to the factor $\sqrt{W}$ in (5.76). We therefore approximate $\sqrt{W} \approx \sqrt{\zeta}$ and take this factor outside the integral for the second and higher terms only. Taking the second term only, it is straightforward to show that the field emission current density j is given approximately by

$$j \approx \frac{e}{2\pi h} \frac{\zeta^{1/2} F^2}{(\zeta + \phi) \phi^{1/2}} \exp \left(-\frac{4\phi^{3/2}}{3F} \sqrt{\frac{2m}{\hbar^2}} \right), \tag{5.81}$$

where we have made use of the definition of the work function ϕ as

$$\phi = C - \zeta, \tag{5.82}$$

and ζ is the Fermi energy. As a reminder, F is the electron charge e times the electric field in volts per meter. It has units of energy per unit length, or joules per meter in this notation. The equation (5.76) and its approximation (5.81) represent the main results of this section. The approximation is precisely the result given by Fowler and Nordheim [31]. It is left as an exercise for the reader to complete the details of this derivation.

Problems

1. Complete the details of the derivation of (5.81).

2. The work function for tungsten is 4.5 electron-Volts. Estimate the field F required for the onset of field emission from tungsten. Describe the functional dependence of the current density j on F for F higher and lower than this onset value.

5.5 Emission with elevated temperature and field

In the preceding sections we explored thermionic emission, and separately cold field emission. In this section we generalize the preceding concepts to calculate the emission current density as a function of temperature *and* applied electric field. We follow the general approach of Murphy and Good [64], which is based on earlier work by Kemble [52].

Again we make use of (5.1) for the emission current density j, and (5.18) for the incident current density per unit energy $J(W)$. The present task is to calculate the transmission probability $D(W)$ that an electron with total energy W in one dimension will tunnel through the potential barrier.

The potential energy $U(x)$ is given by (5.38), and is plotted as a function of x in Figure 5.3. We intentionally include the image potential term in the following. The potential energy $U(x)$ is assumed to join smoothly on both sides of the emission surface at $x = 0$. In the vacuum $(x > 0)$, the potential energy (5.38) has a

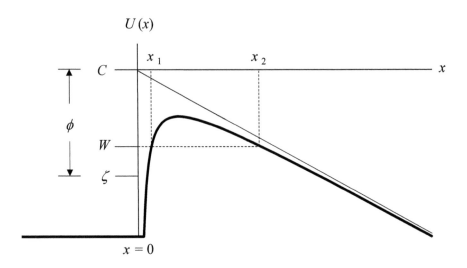

Figure 5.3: Energy diagram for emission with elevated temperature and field.

maximum value U_m given by

$$U_m = C - \sqrt{\frac{e^2 F}{4\pi\epsilon_0}}. \tag{5.83}$$

The total energy W of a single electron inside the material ($x < 0$) can now take on any nonnegative value $0 \le W < \infty$ depending on the absolute temperature T. For $0 \le W \le U_m$ quantum mechanical tunneling occurs. For $W \ge U_m$ no tunneling occurs, but the elevated field F and temperature T act together to enhance the emission.

The spatial part of the wave function $u(x)$ satisfies Schrödinger's equation, which can be expressed in the form

$$\frac{d^2}{dx^2} u(x) + \frac{[p(x)]^2}{\hbar^2} u(x) = 0, \tag{5.84}$$

where $p(x)$ is the kinetic momentum given in one dimension by

$$p(x) = \pm\sqrt{2m[W - U(x)]}. \tag{5.85}$$

We choose the positive root without loss of generality.

All relevant information is contained in the wave function $u(x)$. We *assume* without loss of generality that the wave function can be expressed as

$$u(x) = \exp\left[\frac{i}{\hbar} S(x)\right],$$ (5.86)

where the function $S(x)$ has yet to be determined. Substituting into Schrödinger's equation, we find after some algebra that

$$\left(\frac{dS}{dx}\right)^2 - i\hbar \left(\frac{d^2S}{dx^2}\right) - p^2 = 0.$$ (5.87)

In the classical limit where the term in $\hbar$ can be ignored, this reduces to the Hamilton-Jacobi equation in one spatial dimension, where the electromagnetic potentials have no explicit time dependence. We therefore identify $S(x)$ with Hamilton's characteristic function.

As before we expand $S(x)$ in a series with powers of $\hbar$ as

$$S(x) = S_0(x) + \hbar\, S_1(x) + \hbar^2\, S_2(x) + \dots.$$ (5.88)

Substituting and collecting terms in the powers of $\hbar$, we find

$$(\, S_0'^2 - p^2\,) + \hbar\,(\, -iS''_0 + 2S'_0 S'_1\,) + \dots = 0.$$ (5.89)

Considering $\hbar$ to be small but variable, the quantities within each of the parentheses must vanish separately. The first equation reduces to

$$\frac{dS_0}{dx} = \pm p(x),$$ (5.90)

where $p(x)$ is the kinetic momentum given above. Integrating between any two coordinates x_0 and x we find

$$S_0(x) = S_0(x_0) \pm \int_{x_0}^{x} p(\xi)\, d\xi.$$ (5.91)

Substituting the second bracket in the series for $S(x)$ we find

$$\frac{dS_1}{dx} = \frac{i}{2p}\frac{dp}{dx},$$ (5.92)

where we have made use of

$$S''_0(x) = \pm p'(x). \tag{5.93}$$

Integrating between the limits x_0 and x we find

$$S_1(x) = S_1(x_0) + i \ln \left[\frac{p(x)}{p(x_0)} \right]^{1/2}. \tag{5.94}$$

Substituting above, and ignoring terms in $\hbar^2$ and higher, we find the approximate solution for the wave function $u(x)$ as

$$u(x) \approx u(x_0) \left[\frac{p(x_0)}{p(x)} \right]^{1/2} \exp \left[\pm \frac{i}{\hbar} \int_{x_0}^{x} p(\xi)\, d\xi \right]. \tag{5.95}$$

This solution for the wave function $u(x)$ represents the WKB approximation in one spatial dimension. This approximation was originally due to Wentzel, Kramers, and Brioullon. It is described in many books on quantum mechanics [79, 59]. The solution breaks down at the classical turning points where the momentum $p(x) = 0$, and is only a valid approximation at points remote from the turning points.

We now consider the case where $0 \leq W \leq U_m$ where tunneling occurs. The quantity $p(x)$ is imaginary for $x_1 \leq x \leq x_2$, where $U(x) \geq W$, and real everywhere else. The wave function $u(x)$ is approximated by

$$
\begin{aligned}
u_+(x < x_1) &= \frac{a_+}{p^{1/2}} e^{iw(x)} \\
u_-(x < x_1) &= \frac{a_-}{p^{1/2}} e^{-iw(x)} \\
u_+(x > x_2) &= \frac{b_+}{p^{1/2}} e^{iw(x)},
\end{aligned} \tag{5.96}
$$

where $w(x)$ is defined as

$$w(x) \equiv \frac{1}{\hbar} \int_{x_0}^{x} p(\xi)\, d\xi. \tag{5.97}$$

This integral applies for *any* lower limit x_0. In the following we choose x_0 far to the left of the barrier. The solution $u_+(x < x_1)$ is the right-propagating incident wave, $u_-(x < x_1)$ is the left-propagating reflected wave, and $u_+(x > x_2)$ is the right-propagating transmitted wave. While this approximation breaks down at the classical turning points x_1 and x_2, the function $w(x)$ is well-behaved everywhere.

In order to calculate the transmission probability $D(W)$, we must calculate the probability current j for the incident, reflected, and transmitted waves. In general

$$j = \frac{i\hbar}{2m} \left[u(x)\, \bar{u}'(x) - \bar{u}(x)\, u'(x) \right]. \tag{5.98}$$

We notice in (5.96) that $u_+(x) = \bar{u}_-(x)$ apart from constants. The current j is thus proportional to the conserved Wronskian, and is therefore conserved with respect to the coordinate x as required.

Far from the classical turning points, and apart from constants, it is easily shown in this approximation that

$$
\begin{aligned}
u &= p^{-1/2}\, e^{iw} \\
\bar{u} &= \bar{p}^{-1/2}\, e^{-i\bar{w}} \\
u' &= \left[\frac{m}{2} p^{-5/2} U' + \frac{i}{\hbar} p^{1/2} \right] e^{iw} \\
\bar{u}' &= \left[\frac{m}{2} \bar{p}^{-5/2} U' - \frac{i}{\hbar} \bar{p}^{1/2} \right] e^{-i\bar{w}},
\end{aligned} \tag{5.99}
$$

where we notice that p and w are both either pure real or pure imaginary. After some algebra we arrive at a general expression for the probability current j as

$$j = \frac{p + \bar{p}}{2m\,|p|}\, e^{i(w-\bar{w})}. \tag{5.100}$$

This is identical with the result obtained by Kemble [52]. Substituting, we arrive at the probability current as follows:

$$j_+(x < x_1) = \frac{|a_+|^2}{m}$$

$$j_-(x < x_1) = -\frac{|a_-|^2}{m}$$

$$j_+(x > x_2) = \frac{|b_+|^2}{m} \exp\left(-\frac{2}{\hbar}\int_{x_1}^{x_2} |p(\xi)|\,d\xi\right). \quad (5.101)$$

In the third of these we have made use of

$$w(x) = \frac{1}{\hbar}\left[\int_{x_0}^{x_1} + \int_{x_1}^{x_2} + \int_{x_2}^{x}\right] p(\xi)\,d\xi \quad (5.102)$$

in the region $x > x_2$, with only the second integral leading to a nonzero contribution to $j_+(x > x_2)$.

The right-propagating solution $u_+(x)$ must connect on both sides of the barrier. To ensure this we take

$$|a_+|^2 = |b_+|^2 \quad (5.103)$$

in the above. The probability that a single electron with total energy W tunnels through the barrier is given by

$$D(W) = \frac{j_+(x > x_2)}{j_+(x < x_1)}. \quad (5.104)$$

This leads immediately to

$$D(W) = \exp\left[-\frac{2}{\hbar}\int_{x_1}^{x_2} |p(x)|\,dx\right] \quad (5.105)$$

in the present WKB approximation. Conservation of total current requires that

$$j_+(x < x_1) + j_-(x < x_1) = j_+(x < x_1). \quad (5.106)$$

Dividing both sides by $j_+(x < x_1)$, it follows that the probability that a single electron with total energy W is reflected by the barrier is $1 - D(W)$.

Substituting for $p(x)$, we write

$$D(W) = \exp\left[-\frac{2\sqrt{2m}}{\hbar}\int_{x_1}^{x_2} \sqrt{C - Fx - \kappa x^{-1} - W}\,dx\right], \quad (5.107)$$

where we have defined a constant

$$\kappa = \frac{e^2}{16\pi\epsilon_0}. \qquad (5.108)$$

This applies only to the tunneling case, where the total energy W is less than U_m. The limits x_1 and x_2 are the classical turning points where

$$U(x_i) = W. \qquad (5.109)$$

This leads to a quadratic equation with two roots

$$x_i = \frac{C - W}{2F} \left[1 \pm \left(1 - \frac{4\kappa F}{(C - W)^2} \right)^{1/2} \right]. \qquad (5.110)$$

Setting $\kappa = 0$, we reproduce the tunneling probability for the wedge-shaped barrier (5.75), with the difference that the leading proportionality constant is set equal to unity. This is due to the fact that the present approach relies on the WKB approximation, whereas the earlier calculation is exact.

We now define two new quantities

$$\begin{aligned} \rho &= \frac{2F}{C - W} x \\ y &= \frac{2\sqrt{\kappa F}}{C - W}. \end{aligned} \qquad (5.111)$$

Substituting and performing some algebra we find

$$D(W) = \exp\left[-\frac{2i\sqrt{2m}\kappa^{3/4}}{\hbar F^{1/4} y^{3/2}} \int_{1-\sqrt{1-y^2}}^{1+\sqrt{1-y^2}} \left(\rho - 2 + y^2 \rho^{-1} \right)^{1/2} d\rho \right]. \qquad (5.112)$$

Following Murphy and Good [64] we define a function $v(y)$ as

$$v(y) = -\frac{3i}{4\sqrt{2}} \int_{1-\sqrt{1-y^2}}^{1+\sqrt{1-y^2}} \left(\rho - 2 + y^2 \rho^{-1} \right)^{1/2} d\rho. \qquad (5.113)$$

This function can be expressed in terms of standard elliptic integrals. The reader is referred to a recent paper by Deane et. al. [22]

for a detailed and current discussion. The transmission probability
$D(W)$ is

$$D(W) = \exp\left[\frac{16\,m^{1/2}\kappa^{3/4}}{3\,\hbar F^{1/4}y^{3/2}}\,v(y)\right] \qquad (5.114)$$

for energies $0 \le W \le W_m$. The transmission probability for energies $W \ge W_m$ is $D(W) = 1$, which the reader can easily verify by
applying the above procedure.

We are now in a position to calculate the emission current density
j for the general case of elevated temperature and field. From (5.1,
5.18, 5.105) we find

$$\begin{aligned}
j \;=\; & \frac{4\pi mekT}{h^3}\int_0^{W_m} dW\,\ln\left[\exp\left(\frac{\zeta - W}{kT}\right)+1\right]\\
& \cdot \exp\left[\frac{16\,m^{1/2}\kappa^{3/4}}{3\,\hbar F^{1/4}y^{3/2}}\,v(y)\right]\\
& +\frac{4\pi mekT}{h^3}\int_{W_m}^{\infty} dW\,\ln\left[\exp\left(\frac{\zeta - W}{kT}\right)+1\right],
\end{aligned}$$

$$(5.115)$$

where the constant y is defined in (5.108, 5.111), and the function
$v(y)$ is defined in (5.113). This represents the main result of this
section.

Problems

1. Calculate the transmission probability $D(W)$ for the wedge-
shaped barrier using the WKB approximation. Compare this result with (5.75).

2. Calculate the exact transmission and reflection probabilities
for a square barrier of height U_0 and width $2a$ for the two cases
$W \ge U_0$ and $W \le U_0$.

3. Calculate the transmission and reflection probabilities for a
square barrier of height U_0 for the two cases $W \ge U_0$ and $W \le U_0$

using the WKB approximation.

4. The present analysis assumes two linearly independent eigen-functions $u_\pm(x)$. The eigenfunction $u_+(x)$ represents a wave that is everywhere right-propagating. The eigenfunction $u_-(x)$ represents a wave that is everywhere left-propagating. This leads to the form (5.105) for the transmission probability $D(W)$. An earlier formulation by Kemble [52] assumes a different pair of linearly in-dependent eigenfunctions $f_u(x)$ and $f_v(x)$. The eigenfunction $f_u(x)$ represents a wave that is left-propagating to the left of the bar-rier (reflected wave), and right-propagating to the right of the barrier (transmitted wave). The eigenfunction $f_v(x)$ represents a wave that is right-propagating to the left of the barrier (incident wave), and left-propagating to the right of the barrier (no wave). Show that this leads to an alternative form for the transmission probability $D(W)$ given by

$$D(W) = \left\{ 1 + \exp\left[\frac{2}{\hbar} \int_{x_1}^{x_2} |p(x)|\, dx \right] \right\}^{-1}. \qquad (5.116)$$

(Hint: Write down the analog of the connection formula (5.109) relating the coefficients of the eigenfunctions for the reflected and transmitted waves. This form for $D(W)$ was assumed by Murphy and Good [64].)

5.6 Space charge limited emission

Emission of charged particles gives rise to a space charge cloud in front of the emission surface. We now investigate the condition where the space charge is sufficiently high to suppress the emission. We imagine two parallel plates of infinite extent, separated by a distance s. The emission surface is at zero potential, and the accelerating anode is at potential U_a. We wish to find an expression for the current density j in the space between the plates, as a function of the accelerating potential U_a and the spacing s. This is given by

$$j = \rho(x)\, v(x), \tag{5.117}$$

where ρ is the space charge density, and v is the particle speed. Charge conservation dictates that the current density j is independent of x. The electrostatic potential $U(x)$ is governed by Poisson's equation, which is given in one dimension as

$$\frac{d^2}{dx^2} U(x) = -\frac{\rho(x)}{\epsilon_0}. \tag{5.118}$$

The particle speed is given by energy conservation as

$$v = \sqrt{\frac{2\, e\, U(x)}{m}}. \tag{5.119}$$

Substituting, we obtain a differential equation for the potential $U(x)$ as

$$U''(x) = \frac{\alpha}{\sqrt{U(x)}}, \tag{5.120}$$

where we have defined the constant α as

$$\alpha \equiv -\frac{j}{\epsilon_0}\sqrt{\frac{m}{2\,e}}. \tag{5.121}$$

We now make use of the fact that

$$\frac{d}{dx}\left(U'^{\,2}\right) = 2\,U'\,U'', \qquad \frac{d}{dx} = U'\,\frac{d}{dU}. \tag{5.122}$$

This yields the differential equation

$$d\left(U'^{\,2}\right) = 2\,\alpha\,\frac{dU}{\sqrt{U}}. \tag{5.123}$$

This is integrated immediately to yield

$$U'^{\,2} = 4\,\alpha\,\sqrt{U} + const. \tag{5.124}$$

At this point we invoke the condition that the field is zero at the emission surface at cutoff. Mathematically, this equilibrium is expressed as $U'(0) = 0$. From before we also have $U(0) = 0$, in which case, the integration constant is zero. Taking the square root of both sides, we obtain

$$U^{-1/4} \, dU = 2 \sqrt{\alpha} \, dx. \tag{5.125}$$

Integrating the left side between the limits $U = 0$ and $U = U_a$, and integrating the right side between the limits $x = 0$ and $x = s$,

$$\frac{4}{3} U_a^{3/4} = 2 \sqrt{\alpha} \, s. \tag{5.126}$$

Squaring both sides, substituting for α, and rearranging factors, we obtain the desired expression for the magnitude of the current density j as

$$j = \frac{4 \, \epsilon_0}{9} \sqrt{\frac{2 \, e}{m} \frac{U_a^{3/2}}{s^2}}. \tag{5.127}$$

This is known as the Child–Langmuir equation for space charge limited emission.

Appendix A

The Fourier transform

As a mathematical method, the Fourier transform provides a powerful, simplifying tool for a variety of physical problems. This derives from the fact that a Fourier transform of a function represents the spectral density of the function in the frequency domain. It is a special case in the general theory of Hilbert spaces. Rather than attempt a complete description of this theory, we will confine our attention here only to those aspects that are directly applicable to the present study.

We consider an arbitrary complex function $f(x)$, defined over the range $-\infty < x < +\infty$. We define the Fourier transform $\tilde{f}(k)$ as

$$\tilde{f}(k) = \int_{-\infty}^{\infty} dx \, e^{-ikx} f(x), \qquad (A.1)$$

where k is called the transform variable, and in general $-\infty < k < +\infty$. We assume that the function $f(x)$ is such that the integral is finite. This is true for most problems of physical interest, where f is well-behaved in this sense. Operating on both sides from the left by

$$\frac{1}{2\pi} \int_{-\infty}^{\infty} dk \, e^{ikx'}, \qquad (A.2)$$

we obtain

$$\frac{1}{2\pi} \int_{-\infty}^{\infty} dk \, e^{ikx'} \, \tilde{f}(k) = \int_{-\infty}^{\infty} dx \, f(x) \left[\frac{1}{2\pi} \int_{-\infty}^{\infty} dk \, e^{-ik(x-x')} \right], \qquad (A.3)$$

where we have reversed the order of integrations on the right side. We recognize the large bracket as an integral representation of the Dirac delta function, namely

$$\frac{1}{2\pi} \int_{-\infty}^{\infty} dk\, e^{-ik(x-x')} = \delta(x - x'), \tag{A.4}$$

where, for a certain broad class of well-behaved functions $f(x)$, this has the property

$$\int_{-\infty}^{\infty} dx\, f(x)\, \delta(x - x') = f(x'). \tag{A.5}$$

It follows that

$$f(x) = \frac{1}{2\pi} \int_{-\infty}^{\infty} dk\, \tilde{f}(k)\, e^{ikx}. \tag{A.6}$$

Evidently, this represents the inverse Fourier transform, as it reproduces the original function $f(x)$. The transform (A.1) together with its inverse (A.6) thus form an intimately related pair. Because $\tilde{f}(k)$ multiplies the phase factor on the right side, it represents the spectral density of $f(x)$ with respect to the frequency k, where k has the dimensions x^{-1}. Evaluating the transform at zero argument, it follows immediately that

$$\tilde{f}(0) = \int_{-\infty}^{\infty} dx\, f(x). \tag{A.7}$$

Evaluating the transform $\tilde{f}$ at zero argument gives the integral of the function f over its whole range. This property will turn out to be very useful.

We now derive several other useful properties. We define the convolution $h(x)$ of two functions $f(x)$ and $g(x)$ as the integral

$$h(x) = \int_{-\infty}^{\infty} dx'\, f(x')\, g(x - x'). \tag{A.8}$$

This operation is often abbreviated by

$$h(x) = f(x) * g(x). \tag{A.9}$$

Substituting the inverse transforms for f and g, we find

$$
\begin{aligned}
h(x) &= \int_{-\infty}^{\infty} dx' \left[\frac{1}{2\pi} \int_{-\infty}^{\infty} dk' \, \tilde{f}(k') \, e^{ik'x'} \right] \\
&\cdot \left[\frac{1}{2\pi} \int_{-\infty}^{\infty} dk \, \tilde{g}(k) \, e^{ik(x-x')} \right].
\end{aligned} \tag{A.10}
$$

Interchanging the order of integrations, this becomes

$$
h(x) = \frac{1}{2\pi} \int_{-\infty}^{\infty} dk \, \tilde{g}(k) \, e^{ikx} \int_{-\infty}^{\infty} dk' \, \tilde{f}(k') \cdot \left[\frac{1}{2\pi} \int_{-\infty}^{\infty} dx' \, e^{i(k'-k)x'} \right]. \tag{A.11}
$$

We recognize the quantity in square brackets as an integral representation of the Dirac delta function $\delta(k' - k)$. This leads to

$$
h(x) = \frac{1}{2\pi} \int_{-\infty}^{\infty} dk \, \tilde{f}(k) \, \tilde{g}(k) \, e^{ikx}. \tag{A.12}
$$

From the definition of the inverse transform $h(x)$ this immediately yields

$$
\tilde{h}(k) = \tilde{f}(k) \, \tilde{g}(k). \tag{A.13}
$$

In words, the Fourier transform of a convolution of two functions is equal to the product of the Fourier transforms of the two functions. This general result is called the convolution theorem.

Next we define the autocorrelation function $F(x)$ of a function $f(x)$ as the integral

$$
F(x) = \int_{-\infty}^{\infty} dx' \, f(x') \, f^*(x' - x), \tag{A.14}
$$

where f^* denotes the complex conjugate of f. Substituting the inverse transforms for f and f^*, we find

$$
\begin{aligned}
F(x) &= \int_{-\infty}^{\infty} dx' \left[\frac{1}{2\pi} \int_{-\infty}^{\infty} dk' \, \tilde{f}(k') \, e^{ik'x'} \right] \\
&\cdot \left[\frac{1}{2\pi} \int_{-\infty}^{\infty} dk \, \tilde{f}^*(k) \, e^{-ik(x'-x)} \right].
\end{aligned} \tag{A.15}
$$

Interchanging the order of integrations, this becomes

$$
F(x) = \frac{1}{2\pi} \int_{-\infty}^{\infty} dk \, \tilde{f}^*(k) \, e^{ikx} \int_{-\infty}^{\infty} dk' \, \tilde{f}(k') \cdot \left[\frac{1}{2\pi} \int_{-\infty}^{\infty} dx' \, e^{i(k'-k)x'} \right]. \tag{A.16}
$$

We recognize the quantity in square brackets as an integral representation of the Dirac delta function $\delta(k' - k)$. This leads to

$$F(x) = \frac{1}{2\pi} \int_{-\infty}^{\infty} dk \, \tilde{f}^*(k) \, \tilde{f}(k) \, e^{ikx}. \qquad (A.17)$$

From the definition of the inverse transform $F(x)$ this immediately yields

$$\tilde{F}(k) = |\tilde{f}(k)|^2. \qquad (A.18)$$

In words, the Fourier transform of the autocorrelation is equal to the absolute square of the transform of f. This general result is called the autocorrelation theorem.

Next we investigate the integral

$$
\int_{-\infty}^{\infty} dx \, |f(x)|^2 = \int_{-\infty}^{\infty} dx \left[\frac{1}{2\pi} \int_{-\infty}^{\infty} dk \, \tilde{f}(k) \, e^{ikx} \right]
$$
$$
\cdot \left[\frac{1}{2\pi} \int_{-\infty}^{\infty} dk' \, \tilde{f}^*(k') \, e^{-ik'x} \right], \qquad (A.19)
$$

where we have substituted the inverse transforms of f and f^* on the right side. Interchanging the order of integrations we obtain

$$
\int_{-\infty}^{\infty} dx \, |f(x)|^2 = \frac{1}{2\pi} \int_{-\infty}^{\infty} dk \, \tilde{f}(k) \int_{-\infty}^{\infty} dk' \, \tilde{f}^*(k')
$$
$$
\cdot \left[\frac{1}{2\pi} \int_{-\infty}^{\infty} dx \, e^{-i(k'-k)x} \right]. \qquad (A.20)
$$

Again recognizing the square bracket as $\delta(k' - k)$, we immediately obtain

$$
\int_{-\infty}^{\infty} dx \, |f(x)|^2 = \frac{1}{2\pi} \int_{-\infty}^{\infty} dk \, |\tilde{f}(k)|^2. \qquad (A.21)
$$

This result is known as Parseval's theorem.

All of the preceding results for one spatial dimension can directly be generalized to two dimensions. For a function $f(x, y)$ defined in two Cartesian dimensions, we define the Fourier transform as

$$
\tilde{f}(k_x, k_y) = \int_{-\infty}^{\infty} dx \int_{-\infty}^{\infty} dy \, e^{-i(k_x x + k_y y)} \, f(x, y). \qquad (A.22)
$$

Operating on both sides from the left by

$$\frac{1}{(2\pi)^2} \int_{-\infty}^{\infty} dk_x \int_{-\infty}^{\infty} dk_y \, e^{i(k_x x' + k_y y')}, \tag{A.23}$$

we obtain, reversing the order of integrations

$$\begin{aligned}
&\frac{1}{(2\pi)^2} \int_{-\infty}^{\infty} dk_x \int_{-\infty}^{\infty} dk_y \, e^{i(k_x x' + k_y y')} \, \tilde{f}(k_x, k_y) \\
={} &\int_{-\infty}^{\infty} dx \int_{-\infty}^{\infty} dy \, f(x, y) \left[\frac{1}{2\pi} \int_{-\infty}^{\infty} dk_x \, e^{-ik_x(x-x')} \right] \\
&\cdot \left[\frac{1}{2\pi} \int_{-\infty}^{\infty} dk_y \, e^{-ik_y(y-y')} \right] \\
={} &f(x', y'), \tag{A.24}
\end{aligned}$$

where we again have made use of the Dirac delta function. We thus obtain

$$f(x, y) = \frac{1}{(2\pi)^2} \int_{-\infty}^{\infty} dk_x \int_{-\infty}^{\infty} dk_y \, e^{i(k_x x + k_y y)} \, \tilde{f}(k_x, k_y). \tag{A.25}$$

This represents the inverse Fourier transform in two Cartesian dimensions.

We now investigate what happens when we set one of the transform variable components k_y equal to zero,

$$\tilde{f}(k_x, 0) = \int_{-\infty}^{\infty} dx \, e^{-ik_x x} \int_{-\infty}^{\infty} dy \, f(x, y). \tag{A.26}$$

We define the projection $f_p(x)$ by integrating over one coordinate as follows:

$$f_p(x) = \int_{-\infty}^{\infty} dy \, f(x, y), \tag{A.27}$$

from which it follows that

$$\tilde{f}(k_x, 0) = \tilde{f}_p(k_x). \tag{A.28}$$

In words, setting one component of the transform variable to zero is equivalent to integrating over that degree of freedom in direct space (x, y). This will turn out to be very useful in reducing the number of degrees of freedom in problems of multiple variables. In particular, it greatly simplifies the problem of the stochastic Coulomb interaction in a charged particle beam.

Continuing this process, it is straightforward to show that

$$\tilde{f}(0,0) = \int_{-\infty}^{\infty} dx \int_{-\infty}^{\infty} dy\, f(x,y). \qquad (A.29)$$

As in the case of one dimension, the transform $\tilde{f}(k_x, k_y)$, evaluated at zero argument, represents the integral of the function $f(x,y)$ over the entire direct space (x, y).

Next, we consider the special case where $f(x, y)$ is a function only of $r = \sqrt{x^2 + y^2}$, and is independent of azimuthal angle ϕ. The two-dimensional Fourier transform is

$$\tilde{f}(k_x, k_y) = \int_0^{\infty} dr\, r\, f(r) \int_0^{2\pi} d\phi\, e^{-ikr\cos\phi}, \qquad (A.30)$$

where $k = \sqrt{k_x^2 + k_y^2}$ is the magnitude of the two-vector $\mathbf{k}$. Here we have expressed the element of area $dx\, dy = r\, dr\, d\phi$ in polar coordinates. This reduces to

$$\tilde{f}(k) = 2\pi \int_0^{\infty} dr\, r\, f(r)\, J_0(kr), \qquad (A.31)$$

where J_0 is the zero order Bessel function, for which an integral representation is given by

$$J_0(x) = \frac{1}{2\pi} \int_0^{2\pi} d\phi\, e^{\pm ix\cos\phi}. \qquad (A.32)$$

The above transform is often referred to as a Bessel transform. The transform $\tilde{f}$ depends only on the magnitude of k. Following the same procedure, the inverse transform is readily found to be

$$f(r) = \frac{1}{2\pi} \int_0^{\infty} dk\, k\, \tilde{f}(k)\, J_0(kr). \qquad (A.33)$$

Thus, the radial symmetry of f leads to a simplification of the Fourier transform and its inverse transform.

As an example, we consider the radially symmetric function

$$f(r) = \frac{a^2}{\pi (r^2 + a^2)^2}.$$
(A.34)

Integrating over area, we find that $f(r)$ is normalized to unity, where

$$2\pi \int_0^\infty dr\, r\, f(r) = 1.$$
(A.35)

The Bessel transform is given by

$$\tilde{f}(k) = \frac{ka}{2\pi} K_1(ka),$$
(A.36)

where K_1 is the modified Bessel function. Here we have made use of the integral form

$$\int_0^\infty \frac{J_\nu(bx)\, x^{\nu+1}\, dx}{(x^2 + a^2)^{\mu+1}} = \frac{a^{\nu-\mu}\, b^\mu}{2^\mu\, \Gamma(\mu+1)} K_{\nu-\mu}(ab)$$
(A.37)

for the special case where $\nu = 0$ and $\mu = 1$, where Γ is the gamma-function.

Projecting $f(r)$ onto one Cartesian axis, the x-axis, we form the function

$$
\begin{aligned}
f_p(x) &= \int_{-\infty}^\infty dy\, f\left(\sqrt{x^2 + y^2}\right) \\
&= \frac{a^2}{\pi} \int_{-\infty}^\infty \frac{dy}{[y^2 + (x^2 + a^2)]^2} \\
&= \frac{a^2}{2(x^2 + a^2)^{3/2}},
\end{aligned}
$$
(A.38)

where we have made use of the form

$$\int_{-\infty}^\infty \frac{dy}{(y^2 + a^2)^2} = \frac{\pi}{2a^3}.$$
(A.39)

The one-dimensional Fourier transform is given by

$$\tilde{f}_p(k_x) = \frac{k_x a}{2\pi} K_1(k_x a).$$
(A.40)

These seemingly esoteric relationships are very useful in the theory of small angle plural scattering.

The above arguments are easily extended to n dimensions, in which case the Fourier transform is defined as

$$\tilde{f}(\mathbf{k}) = \int d^n \mathbf{x}\, f(\mathbf{x})\, e^{-i\mathbf{k}\cdot\mathbf{x}}, \qquad (A.41)$$

where $\mathbf{x}$ and $\mathbf{k}$ are n-vectors, and $\mathbf{k}\cdot\mathbf{x} = k_1 x_1 + \ldots + k_n x_n$ is the inner product. Applying the preceding logic, the inverse Fourier transform is found to be

$$f(\mathbf{x}) = \frac{1}{(2\pi)^n} \int d^n \mathbf{k}\, \tilde{f}(\mathbf{k})\, e^{i\mathbf{k}\cdot\mathbf{x}}. \qquad (A.42)$$

The convolution theorem in n dimensions is found to be

$$\tilde{h}(\mathbf{k}) = \tilde{f}(\mathbf{k})\, \tilde{g}(\mathbf{k}), \qquad (A.43)$$

where the n-dimensional convolution is defined as

$$h(\mathbf{x}) = \int d^n \mathbf{x}'\, f(\mathbf{x}')\, g(\mathbf{x} - \mathbf{x}'), \qquad (A.44)$$

and the integration is performed over all of space. The autocorrelation theorem in n dimensions is found to be

$$\tilde{F}(\mathbf{k}) = |\tilde{f}(\mathbf{k})|^2, \qquad (A.45)$$

where the n-dimensional autocorrelation function is defined as

$$F(\mathbf{x}) = \int d^n \mathbf{x}'\, f(\mathbf{x}')\, f^*(\mathbf{x}' - \mathbf{x}). \qquad (A.46)$$

Parseval's theorem in n dimensions is found to be

$$\int d^n \mathbf{x}\, |f(\mathbf{x})|^2 = \frac{1}{(2\pi)^n} \int d^n \mathbf{k}\, |\tilde{f}(\mathbf{k})|^2. \qquad (A.47)$$

This gives us all of the necessary tools to apply the powerful formalism of Fourier analysis to practical problems of charged particle optics.

Appendix B

Linear second-order differential equation

The paraxial ray equation (2.138, 2.162, 2.239) are examples of a more general linear second-order differential equation. We seek a solution for a function $y(x)$, which satisfies an equation of the general form

$$P(x)\frac{d^2y}{dx^2} + Q(x)\frac{dy}{dx} + R(x)\,y = S(x), \qquad (B.1)$$

where P, Q, R, and S are known functions of x. We shall see in the following that this applies directly to the problem of the chromatic aberration.

In the case $S = 0$ the equation is designated homogeneous, and in the case $S \neq 0$ the equation is designated inhomogenous. The solution $y_h(x)$ of the homogenous equation can always be expressed as a linear combination of two functions $u_1(x)$ and $u_2(x)$ as follows:

$$y_h(x) = c_1\,u_1(x) + c_2\,u_2(x), \qquad (B.2)$$

where c_1 and c_2 are constants, and where u_1 and u_2 are not constant multiples of one another. The paraxial ray equations (2.162, 2.239) for the transverse displacement $v(z) = x(z) + i\,y(z)$ in the rotated system is an example of just such a homogeneous equation.

We state the problem as follows: given a solution $y_h(x)$ to the homogeneous equation, find a solution to the inhomogeneous equation. A theorem states that the solution to the inhomogeneous equation can always be expressed as the sum of the homogeneous solution, plus *any* particular solution to the inhomogeneous equation, i.e.,

$$y(x) = y_h(x) + y_p(x). \tag{B.3}$$

We now proceed to find a general solution for $y(x)$, given $y_h(x)$.

For the particular solution y_p we postulate a trial function

$$y_p(x) = C_1(x)\, u_1(x) + C_2(x)\, u_2(x), \tag{B.4}$$

where C_1 and C_2 have yet to be specified. In the following we adopt the notation

$$\frac{dy}{dx} = y'(x), \qquad \frac{d^2 y}{dx^2} = y''(x). \tag{B.5}$$

Differentiating (B.4), we find

$$
\begin{aligned}
y_p' &= C_1\, u_1' + C_2\, u_2' + C_1'\, u_1 + C_2'\, u_2 \\
y_p'' &= C_1\, u_1'' + C_2\, u_2'' + 2\, C_1'\, u_1' + 2\, C_2'\, u_2' + C_1''\, u_1 + C_2''\, u_2.
\end{aligned}
\tag{B.6}
$$

We are free to select one arbitrary condition on C_1 and C_2. We choose this to be

$$C_1'\, u_1 + C_2'\, u_2 = 0, \tag{B.7}$$

thus eliminating the last two terms in y_p'. Differentiating (B.7), we find

$$C_1'\, u_1' + C_2'\, u_2' + C_1''\, u_1 + C_2''\, u_2 = 0. \tag{B.8}$$

This reduces y_p'' to

$$y_p'' = C_1\, u_1'' + C_2\, u_2'' + C_1'\, u_1' + C_2'\, u_2'. \tag{B.9}$$

Substituting the reduced y_p' and y_p'' into (B.1), we find

$$C_1'\, u_1' + C_2'\, u_2' = \frac{S}{P}. \tag{B.10}$$

where we assume P is nonzero. Together with (B.7), this gives a pair of simultaneous equations for C_1' and C_2'. Solving this pair, we find

$$C_1'(x) = -\frac{S(x)}{P(x)\,W(x)}\,u_2(x)$$

$$C_2'(x) = \frac{S(x)}{P(x)\,W(x)}\,u_1(x), \tag{B.11}$$

where $W(x)$ is the determinant defined as

$$W(x) = u_1(x)\,u_2'(x) - u_1'(x)\,u_2(x). \tag{B.12}$$

The pair (B.11) can be integrated in principle to give

$$C_1(x) = -\int \frac{S\,u_2}{P\,W}\,dx$$

$$C_2(x) = \int \frac{S\,u_1}{P\,W}\,dx, \tag{B.13}$$

from which it follows (B.2, B.4) that

$$y(x) = \left(-\int \frac{S\,u_2}{P\,W}\,dx + c_1\right)u_1(x) + \left(\int \frac{S\,u_1}{P\,W}\,dx + c_2\right)u_2(x). \tag{B.14}$$

This represents the general solution to (B.1).

We are now in a position to apply this directly to the problem of chromatic aberration in the case of axial symmetry. The inhomogeneous equation for $\delta v_1(z)$ is (2.242). We identify the solution (B.2) to the homogeneous equation with

$$\delta v_{1h}(z) = \delta v_{1O}\,g(z) + \delta v_{1A}\,h(z), \tag{B.15}$$

where $\delta v_{1O} = 0$, as there is no aberration in the object plane. Also, $P = 1$, and

$$W = g\,h' - g'\,h = k\,\mathsf{p}^{-1}(z). \tag{B.16}$$

The inhomogeneous term S is given by (2.243). Substituting these into (B.14), the chromatic aberration in the Gaussian image plane is given by

$$\delta v(z_I) = -\frac{M}{k}\int_{z_O}^{z_I} \mathsf{p}(z)\,S(z)\,h(z)\,dz, \tag{B.17}$$

where we have made use of $g(z_I) = $ M and $h(z_I) = 0$. This is identical with (2.242).

Bibliography

[1] Abramowitz M. and Stegun I.A. (eds.) 1965. *Handbook of Mathematical Functions*, Dover Publications, Inc., New York, ISBN-13: 978-0-486-61272-0.

[2] Aharonov Y. and Bohm D. 1959. Significance of electromagnetic potentials in the quantum theory, *Phys. Rev.*, Second Series, 115(3), 485.

[3] Batelaan H. and Tonomura A. 2009. The Aharonov–Bohm effects: Variations on a subtle theme, *Physics Today*, 62(9), 38.

[4] Beck V.D., Gordon M.S. and Groves T.R. 1989. A fast Monte Carlo simulator using exact Coulomb scattering, *J. Vac. Sci. Technol.* B, 7(6), 1438.

[5] Bethe H. 1930. Zur Theorie des Durchgangs schneller Korpuskularstrahlen durch Materie, *Ann. d. Physik*, Ser. 5, Vol. 5, 325.

[6] Bjorken J.D. and Drell S.D. 1964. *Relativistic Quantum Mechanics*, McGraw Hill, ISBN-13: 978-0070054936.

[7] Boersch H. 1954. *Zeitschr. Physik* 139, 115.

[8] Bohm D.J. 1989. *Quantum Theory*, Dover Press, ISBN-13: 978-0486659695.

[9] Born M. 1926. *Zeitsch. Physik* 37, 863.

[10] Born M. 1927. *Nature* 119, 354.

[11] Born M. and Wolf E. 1999. *Principles of Optics*, Seventh edition, Cambridge University Press. ISBN-13: 978-0521642224.

[12] Brüche E. and Scherzer O. 1934. *Geometrische Elektronenoptik: Grundlagen und Anwendungen*, Berlin.

[13] Brunner T.A. 1997. Impact of lens aberrations on optical lithography, *IBM J. Research Devel.* 41, 12, 57.

[14] Busch H. 1926. *Ann. Physik* 81, 974.

[15] Busch H. 1927. *Arch. Elektrotech.* 18, 583.

[16] Buseck P., Cowley J. and Eyring L. (eds.) 1988. *High-Resolution Transmission Electron Microscopy and Associated Techniques*, Oxford University Press, ISBN 0-19-504275-1.

[17] Chandrasekhar S. 1943. Stochastic problems in physics and astronomy, Chapter I. The problem of random flights, *Rev. Mod. Physics* 15, 1, 1.

[18] Corson D.R. and Lorrain P. 1970. *Introduction to Electromagnetic Fields and Waves*, Second edition, W.H. Freeman and Co., ISBN-13: 978-0716703310.

[19] Crewe A.V., Wall J. and Langmore J. 1970. Visibility of Single Atoms, *Science* 168, 1338.

[20] Crewe A.V. 1973. Production of Electron Probes Using a Field Emission Source, *Progress in Optics* XI, Wolf E. (ed.), North–Holland.

[21] Crewe A.V. and Groves T.R. 1974. Thick Specimens in the CEM and STEM. I. Contrast, *J. Appl. Physics* 45, 8, 3662.

[22] Deane J.H.B., Forbes R.G. and Shail R.W. 2007. Formal derivation of an exact series expansion for the principal field emission elliptic function v, arXiv:0708.0996v2 (math-ph).

[23] Dirac P.A.M. 1982. *The Principles of Quantum Mechanics, International Series of Monographs on Physics* 27, Fourth edition, Oxford University Press, ISBN 0-19-8520011 5.

[24] Egerton R.F. 1996. *Electron Energy-Loss Spectroscopy in the Electron Microscope*, Second edition, Plenum Press, ISBN 0-306-45223-5.

[25] Ehrenberg W. and Siday R.E. 1949. The refractive index in electron optics and the principles of dynamics, *Proc. Physical Soc.* B, 62, 8.

[26] El-Kareh A.B. and Smither M.A. 1979. *J. Appl. Phys.* 50, 5596.

[27] Fermi, E. 1995. *Notes on Quantum Mechanics*, University of Chicago Press, ASIN: B000OPN6HA.

[28] Feynman R.P. 1948. Space-time approach to non-relativistic quantum mechnics, *Rev. Mod. Physics* 20, 2, 367.

[29] Feynman R.P. and Hibbs A.R. 1965. emended by Styer D.F. 2005. *Quantum Mechanics and Path Integrals*, Dover Publications, Inc. 2010. ISBN-13: 978-0-486-47722-0.

[30] Feynman R.P., Leighton R.B. and Sands M. 2011. *The Feynman Lectures on Physics*, Vol. III, Basic Books, ISBN-13: 978-0465025015.

[31] Fowler R.H. and Nordheim L. 1928. Electron emission in intense electric fields, *Proc. Royal Soc. London* A, 119, 781, 173.

[32] Gallatin G.M. 2000. Analytic evaluation of the intensity point spread function, *J. Vacuum Sci. Technol.* B, 18(6), 3023.

[33] Glaser W. 1952. *Grundlagen der Elektronenoptik*, Springer–Verlag Berlin Heidelberg GmbH, ISBN-13: 978-3662236208.

[34] Goldstein J., Newbury D.E., Joy D.C., Lyman C.E., Echlin P., Lifshin E., Sawyer L. and Michael J.R. 2007. *Scanning Electron Microscopy and X-ray Microanalysis*, Third edition, Springer Science and Business Media, Inc., ISBN-13: 978-0306472923.

[35] Goldstein H., Poole C.P. and Safko J.L. 2001. *Classical Mechanics*, Third edition, Addison–Wesley, ISBN-13: 978-0201657029.

[36] Goodman J.W. 2004. *Introduction to Fourier Optics*, Third edition, Roberts and Company Publishers, ISBN-13: 978-0974707723.

[37] Gradshteyn I.S., Ryzhik I.M. and Jeffrey A. 1994. *Table of Integrals, Series, and Products*, Fifth edition, Academic Press, ISBN-13: 978-0122947551.

[38] Groves T.R. 1975. Thick specimens in the CEM and STEM, resolution and image formation, *Ultramicroscopy* 1, 15.

[39] Groves T.R., Hammond D.L. and Kuo H.P. 1979. Electron beam broadening effects caused by discreteness of space charge, *J. Vacuum Sci. Technol.* 16(6), 1680.

[40] Groves T.R. 1994. Efficiency enhancement of Monte Carlo simulation of particle beam interaction by separation of stochastic and continuum contributions, *J. Vacuum Sci. Technol.* B, 12(6), 3483.

[41] Groves T.R. 1999. Theory of Coulomb scattering in particle beams using Markov's method, *J. Vacuum Sci. Technol.* B, 17(6), 2808.

[42] Hall C.E. 1966. *Introduction to Electron Microscopy*, McGraw–Hill, ASIN: B0000CN58A.

[43] Hawkes P.W. and Kasper E. 1996. *Principles of Electron Optics*, Vol. 1, Basic Geometrical Optics, Elsevier (present), Academic Press (original), ISBN 0-12-333341-5.

[44] Hawkes P.W. and Kasper E. 1989. *Principles of Electron Optics*, Vol. 2, Applied Geometrical Optics, Elsevier (present), Academic Press (original), ISBN 0-12-333352-0.

[45] Hawkes P.W. and Kasper E. 1996. *Principles of Electron Optics*, Vol. 3, Wave Optics, Elsevier (present), Academic Press (original), ISBN 0-12-333343-1.

[46] Herrmann G. and Wagener S. 1951. *The Oxide-coated Cathode*, Vol. 2, Physics, Translated by Wagener S., Chapman & Hall, ASIN: B0007J3GMS.

[47] Hopkins R.E. 1962. *Research on Fundamentals of Geometrical Optics*, Pubs: PN, ASIN: B009TH4G0K.

[48] Jackson J.D. 1998. *Classical Electrodynamics*, Third edition, John Wiley & Sons, ISBN-13: 978-0471309321.

[49] Jansen G.H. 1990. *Coulomb interactions in particle beams*, *Adv. Electronics Electron Phys.*, Suppl. 21, Academic Press, ISBN-13: 978-0120145836.

[50] Keil E., Zeitler E. and Zinn W. 1960. Zur Einfach- und Mehrfachstreuung geladener Teilchen, *Zeitsch. Naturforsch.* 15a, 12, 1031.

[51] Kelly J., Groves T.R. and Kuo, H.P. 1981. A high current, high speed electron beam lithography column, *J. Vacuum Sci. Technol.* 19, 4, 936.

[52] Kemble E.C. 1937, 1958, 2005. *The Fundamental Principles of Quantum Mechanics*, Dover Phoenix Editions, Dover Publications, Inc. ISBN 0-486-44153-9.

[53] Klemperer O. and Barnett M.E. 2011. *Electron Optics*, Cambridge Monographs on Physics, Third Edition, Cambridge University Press, ISBN-13: 978-0521179737.

[54] Krivanek O.L., Dellby N. and Murfitt M.F. 2009. Aberration correction in electron microscopy, *Handbook of Charged Particle Optics*, Second edition, Chapter 12, 601–640, Orloff J. (ed.), CRC Press, Taylor & Francis Group, ISBN-13: 978-1-4200-4554-3.

[55] Kruit P. and Jansen G.H. 2009. Space charge and statistical Coulomb effects, *Handbook of Charged Particle Optics*, Second Edition, Chapter 7, 341-389, Orloff J. (ed.), CRC Press, Taylor & Francis Group, ISBN-13: 978-1-4200-4554-3.

[56] Landau L.D. and Lifshitz E.M. 1979. *Electrodynamics of Continuous Media, Course of Theoretical Physics*, Vol. 8, Second edition, Butterworth–Heinemann, ISBN-13: 978-0750626347.

[57] Landau L.D. and Lifshitz E.M. 1981. *Quantum Mechanics, Non-relativistic Theory, Course of Theoretical Physics*, Vol. 3, Third edition, Butterworth–Heinemann, ISBN-13: 978-0080291406.

[58] Levi-Setti R., Gavrilov K.L., Neilly M.E., Strick R., and Strissel P.L. 2005, High resolution SIMS imaging of cations in mammalian cell mitosis, and in Drosophila polytene chromosomes, *Appl. Surface Sci.*, 252, 6907.

[59] Liboff R.L. 2002. *Introductory Quantum Mechanics*, Fourth edition, Addison–Wesley, ISBN-13: 978-0805387148.

[60] Leighton R.B. 1959. *Principles of Modern Physics*, McGraw–Hill Education, ISBN-13: 978-0070371309.

[61] Martin W.T. and Reissner E. 1986. *Elementary Differential Equations*, Reprinting of second edition, Dover Press, ISBN 0486650243.

[62] Mott N.F. and Massey H.S.W. 1987. *The Theory of Atomic Collisions: Vol. I, International Series of Monographs on Physics*, Third edition, Oxford University Press, ISBN-13: 978-0198520306.

[63] Munro's Electron Beam Software, Ltd., http://www.mebs.co.uk/

[64] Murphy E.L. and Good R.H. Jr. 1956. Thermionic emission, field emission, and the transition region, *Phys. Rev.* 102, 6, 1464 - 1473.

[65] Nion, Inc. http://www.nion.com/

[66] Nordheim L.W. 1928. The effect of the image force on the emission and reflexion of electrons by metals, *Proc. Royal Soc. London, Series A* 121, 788, 626.

[67] Orloff, J. (ed.) 2009. *Handbook of Charged Particle Optics*, Second edition, CRC Press, Taylor & Francis Group, ISBN-13: 978-1-4200-4554-3

[68] Pfeiffer H.C. 1971. *Eleventh Symposium on Electron, Ion, and Laser Beam Technology*, 239.

[69] Pfeiffer H.C. 1972. *IIT Symposium on Scanning Electron Microscopy*, 113.

[70] Pfeiffer H.C., Groves T.R. and Newman T.H. 1988. High throughput, high-resolution electron-beam lithography, *IBM J. Research* 32, 4, 494.

[71] Picht J. 1932. *Ann. Physik* 15, 926.

[72] Protter M.H. and Morrey C.B. 1970. *Modern Mathematical Analysis*, Addison–Wesley, ASIN: B003P179UA.

[73] Reimer L. and Kohl H. 2009. *Transmission Electron Microscopy, EJB Reviews* Federation of the European Biochemical Soc., Springer, ISBN-13: 978-0387564142.

[74] Reiser M. 2008. *Theory and Design of Charged Particle Beams*, Wiley Series in Beam Physics and Accelerator Technology, Second edition, Wiley–VCH, ISBN-13: 978-3527407415.

[75] Rose H. 2013. *Geometrical Charged Particle Optics*, Second edition, Springer Series in Optical Sciences, ISBN-13: 978-3642321184.

[76] Sasaki, T. 1979. *Conference on Very Large Scale Integration: Architecture, Design, and Fabrication*, California Institute of Technology.

[77] Scherzer O. 1936. *Zeitschr. Physik* 101, 593.

[78] Scherzer O. 1947. *Optik* 2, 114.

[79] Schiff L.I. 1968. *Quantum Mechanics*, International Series on Pure and Applied Physics, Third edition, McGraw–Hill College, ISBN-13: 978-0070552876.

[80] Scott W.T. 1963. *Rev. Mod. Physics* 35, 231.

[81] Snyder H.S. and Scott W.T. 1949. *Phys. Rev.* 76, 220.

[82] Sommerfeld A. 1928. Zur Elektronentheorie der Metalle auf Grund der Fermischen Statistik, *Zeitschr. Physik* 47, 1.

[83] Sommerfeld A. 1964. *Mechanics, Lectures on Theoretical Physics*, Volume 1, Translated from the Fourth German edition by Martin O. Stern, Academic Press, ISBN-13: 978-0126546705.

[84] Sommerfeld A. 1964. *Electrodynamics, Lectures on Theoretical Physics*, Volume 3, Translated from the Fourth German edition by Martin O. Stern, Academic Press, ISBN-13: 978-0126546644.

[85] Sommerfeld A. 1964. *Optics, Lectures on Theoretical Physics*, Volume 4, Translated from the Fourth German edition by Martin O. Stern, Academic Press, ISBN-13: 978-0126546767.

[86] Sturrock P.A. 1955. *Static and Dynamic Electron Optics*, Cambridge University Press, ASIN: B00DSK1UBW.

[87] Sturrock P.A. and Groves, T.R. 2010. More Variations on Aharonov-Bohm, *Physics Today*, 63(4), 8.

[88] Tang T.T. 1983. Monte Carlo simulation of discrete space charge effects in e-beam lithography systems, *Optik* 64, 3, 237.

[89] Van Leeuwen J.M.J. and Jansen G.H. 1983. *Optik* 65, 179.

[90] Varghese L.T. and Fan L. 2010. *Electron, Ion, and Photon Beam Nanolithgraphy Symposium* (EIPBN) Micrograph Competition, www.eipbn.org.

[91] Walker J. 2011. *Halliday and Resnick, Fundamentals of Physics*, Ninth edition, John Wiley & Sons, ISBN 978-0-470-46908-8 (hardbound).

[92] Williams D.B. and Carter C.B. 1996. *Transmission Electron Microscopy: A Textbook for Materials Science*, Volumes 1–4, Second edition, Plenum Press, ISBN-13: 978-0387765020.

[93] Wollnik H. 1987. *Optics of Charged Particles*, Academic Press, ISBN-13: 978-0124333666.

[94] Zworykin V.K., Morton G.A., Ramberg E.G., Hillier J. and Vance A.W. 1961. *Electron Optics and the Electron Microscope*, Fifth Printing, John Wiley & Sons, New York, ASIN: B0019MV3E0.

Index